# CONDUCTOR

# The Heart and Soul of the Railroad

BY

ROBERT L. BRYAN

This book is dedicated to Marilyn, Bryan, and Meghan, now more than ever the angel on my shoulder

# TABLE OF CONTENTS

# PROLOGUE

As a New York City boy, aside from a set of Lionel trains that I owned as a young child, I had no interest in the railroad as a career or a hobby. During most of my adult life I have lived in the shadow of the Long Island Railroad, the largest commuter railroad in North America. Despite my close proximity to the LIRR, I hardly ever utilized the railroad. Because I stilled lived in the New York City borough of Queens, during my entire career with the New York City Police Department, I either drove to and from work, or utilized the city subway or bus lines.

My interest in the railroad commenced long after my retirement from the NYPD, and had its roots in an unspeakable personal tragedy. On December 19, 2011 I was involved in a vehicular accident in which a driver under the influence of cocaine drove through a red light at a high rate of speed and T-boned my vehicle. My daughter Meghan was a senior at college and I had been driving her to school for one of her final examinations. As a result of the collision my neck and back were broken, all my ribs were broken, most of my teeth were knocked out, my right hand was

5

shattered, and I experienced damage to the peroneal nerve that requires me to wear a brace on my left leg. Simply stated, I was in bad shape. Ultimately, however, I recovered to the point where I could continue normal life activities. My beautiful daughter Meghan, however, was killed.

Once I was home from the hospital and rehabilitation facility, I had many months to recover physically and emotionally. When an average day at home consisted of sitting on the sofa doing nothing, my mind was constantly searching for things to do to stay active and to keep from obsessing on my recent tragedy. My son, Bryan, was finishing up his second year of college. Bryan had not shown any great interest in any particular profession, nor did he seem enthusiastic about following me into a law enforcement career. One of my regular activities during this time period was to surf the Internet in search of potential career opportunities for Bryan. One day I chanced upon the Long Island Railroad job board and noticed that the position of Assistant Conductor was open. As I previously mentioned, I knew very little about the LIRR, but I did know that the position of Conductor was highly sought after because of its good pay and benefits. I was surprised to see that the only formal requirements for the job were a high school diploma, with some college and cash handling experience preferred. Bryan had worked at a supermarket for the prior three years, so I quickly threw together a resume for him that highlighted his two years of college and cash handling experience at the supermarket. I emailed the resume and went on to any other activities that I could find to pass the time. I never even mentioned to Bryan that he had applied for the job.

Several weeks later I received an email, or should I say Bryan received an email inviting him to take the Assistant Conductor test. In retrospect, I believe there was divine intervention on Bryan's behalf because I have subsequently read numerous stories of people who have been applying for LIRR jobs for over a decade, and have never been called for any further consideration. Approximately two weeks later Bryan reported to the LIRR's Hillside Support Facility and completed the process, which included testing in vocabulary, math, a cognitive test, and a preliminary interview. Bryan returned home feeling that he had performed well on the tests, but no one had told him if he had passed or what the next step in the process was going to be.

Four days later I received an email inviting Bryan to the Signals and Definition overview. Even though I had absolutely no idea what a Signals and Definition overview was, I knew this was good news as they would not be calling Bryan back if he had not passed the initial tests. When Bryan returned from the overview he was carrying a booklet that contained numerous railroad definitions and drawings of signals, along with their meanings. Bryan explained that he had a month to study these materials before returning for a test. This did not seem so bad to me, but then Bryan continued that if he got one signal incorrect, he failed. He further stated that all the definitions had to be written out verbatim. I scanned the pages and definitions and noticed that some of the definitions were several sentences in length. I questioned if he had understood the LIRR personnel correctly regarding having to write these lengthy definitions word for word. Bryan insisted that he was correct, and he elaborated that one of the LIRR instructors said that if an answer was 99% correct, it was actually 100% wrong.

Bryan must have realized what a great opportunity he was being given, because he really buckled down and put in the time and effort to learn those signals and definitions to the point where he could spit them out verbatim with no problem. I was a nervous wreck the day he took the test as I waited for his call. I was elated when he called with the news that he had passed. Little did I realize that this was just the beginning of my stressful waiting periods for his calls.

Over the next several weeks Bryan was notified to appear for medical and physical testing, and then there was nothing more to do than wait. Finally, on June 3, 2013 Bryan received a letter offering him appointment as an Assistant Conductor on June 19, 2013. The eleven-week Assistant Conductor training course was intense. During my twenty-year career with the NYPD I experienced all different varieties of specialized training programs, from the recruit training at the police academy to the management and leadership training provided to newly promoted captains. Frankly, I don't know if I could have passed the Assistant Conductor course as Bryan did. The amount of information that had to be memorized was staggering. Every few weeks, the class had to take a test, and if you failed the test….you were out. That's right…fired, terminated, gone. If a police recruit fails the academy, they are placed in a "holdover" program, receive additional training and another opportunity to pass the test.  During the weeks of training I would sit with Bryan

almost every night and drill him with flashcards on the hundreds of

definitions and signals that he had to learn word for word.  For

example, a typical definition was:

*Explain Rule 261*

*On portions of the railroad and on tracks specified in the timetable,*

*trains will be governed by block signals whose indications will*

*supersede the superiority of trains for both opposing and following*

*movements on the same track Trains MUST NOT clear or enter the*

*main track at a switch not equipped with an electric lock without*

*Form L authority.*

Learning the definition of Rule 261 may not appear overly

challenging, but try memorizing the definition word for word along

with several hundred additional definitions.  I still vividly remember

those test days, waiting for the call from Bryan and letting out a huge

sigh of relief when he quickly blurted out that he passed.  As

Bryan's training progressed, a strange thing began to happen.  As he

learned about the railroad and the job of Assistant Conductor, so did

I.  I began to develop a strong interest in the inner workings of the

railroad and the job that Conductors actually perform.  I wanted to know what the abstract terms that I helped Bryan study, like

Absolute-Medium-Clear, Approach-Slow, and Block-limit actually

meant in the day to day life of a conductor. I also became interested

in the different types of trains, or equipment, as they are known to

railroaders, used by the LIRR. What was the difference between M-

3's and M-7's, and what different types of locomotives were used?

As Bryan began to actually work on the trains, I became increasingly

curious about the various roles performed by the Conductor. What

specifically does the Brakeman do? What was the role of a

Collector? Who was actually in charge of the train – the Conductor

or the Engineer?

    Bryan's new career with the railroad had spawned a new

hobby for me. I certainly was not at the point where I was standing

on station platforms taking videos of the various locomotives and

trains, but I had definitely developed an interest in the railroad, and

in particular the position of Conductor. Meanwhile, Bryan's

studying never stopped. Federal Railroad Administration regulations

require that all Conductors be fully qualified, so to progress to the

title of Conductor, all Assistant Conductors have to pass a series of

tests that are far more rigorous that the tests administered during the initial training period. Bryan had been an Assistant Conductor for about three years when he finally completed his qualification. As with the initial training course, those who could not pass a qualification test were ultimately terminated.

My son Bryan is now well into his career with the railroad and I am well into my hobby. I have done a lot of reading on the role of the railroad conductor, and my research is contained in the remainder of this book. I have found the railroad conductor to be a slice of Americana and a fascinating position, and I hope you do too.

## Introduction: Picture the Conductor

"Time to be traveling again," says Ringo Starr in his role as Shining Time Station's Mr. Conductor. He pulls out the chained pocket watch that has become so characteristic of the profession. When he's not telling tales of his adventures as a conductor, he's making sure the trains run on time, one of the many responsibilities entrusted to this classic profession. Whether portrayed by Starr, Carlin, Baldwin, or another, Mr. Conductor is both sage advisor and gatekeeper to a magical world in which sentient trains participate in grand adventures. As a conductor, he is considered an authority on trains and a figure demanding of respect. This was my introduction into the world of conductors as Bryan fell in love with the characters on Shining Time Station as a young child.

Train conductors have taken on a unique role in popular culture as a profession that is at once contemporary and antiquated. Even today, they may be seen wearing the trademark vest, hat, and watch that one might have expected a conductor to wear a hundred years ago. They are romanticized and, in some ways, frozen in time. However, that does not change the fact that these professionals are still fulfilling their duties on trains across the world. The job of a

train conductor is a very real and very important one.

In this book, we will discuss the ins and outs of this storied career,

where it started, how it influenced transportation in North America

and across the globe, and where its future lies. We will look closely

at a profession that symbolizes both professionalism and adventure,

punctuality and freedom.

Classic Conductor
Images

# Chapter One: The Railroad

Without trains there would be no train conductors. This may seem like an obvious statement, and it is, but it also leads us to an important truth. Before we can fully appreciate the vital job of the train conductor, we need to take some time to explore trains, how and why they came about, and the significant role they play in our history and our lives.

Picture the American "Wild West." It's rife with cowboys and saloons. But that's not all. The westward bound railway, complete with workers, conductors, and fancy ladies with parasols, was a fixture of the west and symbol of westward expansion. Sure, it was an easy mode of transportation and an effective way to expand. But let's not underestimate the tremendous power and influence that things which are "easy" have over the course of history.

American railroads made it possible to grow a young country at an unprecedented pace, making it easier to bring goods and people from one end of a continent to the other. While some may argue that the railroad made it easier to push aside native people, no one can deny the awesome influence the railroad had in shaping modern America and the modern world.

In both America and Britain, railroads made it easier to mine goods faster, feeding factories and, therefore, their respective economies. In many ways, the history of the railroad is the history of "easy." Its legacy is the tearing down of barriers in the path of progress and expansion. It was the first step in making an unfathomably large world an intimately connected one.

*The Invention of the Modern Train*

Rail transport refers to the movement of cargo from one location to another on vehicles that run on tracks. These tracks, though limiting the possible routes a vehicle could take, are designed to reduce friction and make the conveyance of goods or people faster and easier. It may surprise you to realize that using tracks or rails to improve transportation has been in practice since several hundred years B.C., evolving over the years to cover greater distances and carry heavier loads.

Many credit wagonways as the earliest railroads. These pathways date back to ancient Greece where they aided in the transport of carts by providing grooves in stone roads within which wheels could run more smoothly. Over time, these wagonways evolved and became raised and wooden. They made it easier to deliver heavy loads and were frequently used to transport fully laden wagons downhill with the help of gravity.

Although these ancient methods of improving transportation definitely stemmed from many of the same scientific principles that modern railroads take advantage of, the railroad as we understand it got its start around the Industrial Revolution. Up until this time, most rails were wooden. The Industrial Revolution inspired the use of iron to reinforce these tracks and, in many cases, eventually replace them.

Toward the end of the Industrial Revolution, the first mechanized rail transport appeared in England. British inventor Thomas Newcomen invented the steam engine, but it was James Watt, a Scottish inventor, who further developed it, creating the first steam powered reciprocating engine. The significance of this can't be overstated. This reciprocating engine, or piston engine, could power a wheel. If we take a moment to appreciate the significant role the wheel has played in the evolution of civilization, we can see how applying a powerful engine to one would quickly become a game changer.

Watt went on to patent the design for a steam locomotive in 1784. However, the first life sized version of the locomotive wouldn't be built until 1804. The design continued to improve over the 19th century, eventually coming to the United States in the late 1820s.

*The Railroad in the United States*

Remember how I asked you to picture the Wild West? Conjure that image again in your mind. The American Revolution may have taken place about a century before, but what you are seeing is the continued birth of a nation. The railroad was so integral to the creation of the United States as we now know it, that the two are largely entwined. And there's no better way to appreciate the significance of the railroad than by looking at the impact it had on America's future.

The Baltimore and Ohio Railroad is among the first ever built in the United States and the first "common carrier." Up until this point, railroads were built for specific purposes by specific companies. They each served the needs of the organization that built them. A common carrier railroad is a railroad that does not serve a single entity. It can be used by any private or public entities who pay for the privilege.

Following the creation of the B and O Railroad and the testing of the first American steam locomotive, the Stourbridge Lion, the United States experienced a railroad heyday. During this time, several lines were laid down in the northeast and some in the south. Railroads quickly became a popular way to travel, replacing canals as the primary mode of transportation in the mid-1800s. This enthusiasm for railroads would continue until about 1873 when a depression swept Europe and North America, ceasing the growth of railroad development.

The depression of the mid to late 1800s was not the only roadblock to further railroad development. American railroads found opposition in a variety of places, not the least of which were the canal and stagecoach companies whose business they threatened. Even taverns and inns decried the building of railroads. They too felt threatened as faster transportation would make for less need to seek their services during long journeys. These critics would site safety reasons among others to dissuade people from using the new method of transportation. But the economic benefits of railroads were undeniable and safety was being seriously addressed.

As railroads developed in the United States, they saw many improvements to make railroad travel safer and more efficient. In the mid-1800s railroads began using electric telegraphs to convey messages. This made it possible to give orders that would adjust train movements, making it easier to keep track of where trains were in order to avoid collisions. In fact, the practice of telegraphy in conjunction with the railroad system was so successful, it continued into the early 20th century. Shortly after the introduction of telegraphers, railroads began implementing manual blocking systems that eventually became automatic block signaling (ABS), a practice still in effect today. This communication method worked by dividing railroads into sections and using signals to control the movement of trains within each section. Doing so not only made the tracks safer, but also made it possible for trains to travel at greater speeds and with greater efficiency.

Despite the lull in railroad development around the mid to late 1800s, building eventually picked up again and the American railroad system was largely complete by the early 20th century. However, by this time, the railroad companies were already finding competition in even newer methods of transportation. Trucks and planes would threaten the usefulness of the railroad system, but, as we've seen today, American railroads would remain alive and well.

*The Transcontinental Railroad*

Although we've already spent some time looking at the overall evolution of the American railroad system, we left out one very key piece of the puzzle. In and around improving transportation and shipment in the east, America was hard at work using railroads to press westward and expand the country's foothold in the western half of the continent.

Prior to the development of the first ever Transcontinental Railroad, traveling out west was a lengthy and dangerous proposition. Whether you decided to hazard the journey by land or attempt to make a shorter trip by sea, you were exposing yourself to disease, hunger, and any number of other perils associated with the great outdoors. It should come as no surprise that there was tremendous interest in building a railroad to unite the young and sprawling country. Congress, therefore, passed the Pacific Railroad Act to fund the building of a railroad that would stretch clear to the Pacific coast. However, it was not the government itself that was to build the tracks. Instead Congress allowed for the issuance of land grants and government bonds to railroad companies. In essence, they wanted the companies to build and own the railroads, they were simply incentivizing them to do so.

The original act, passed in 1862, and its successors resulted in the Union Pacific Route and Central Pacific Route. The Union Pacific Route started in Bluffs, Iowa and ran westward. The Central Pacific Route started in Sacramento, California and ran eastward. The two lines met in Promontory Summit, Utah, creating a single contiguous railroad line across the majority of the United States. Each line was built and controlled by a separate company and they competed throughout the building of the rail lines to see who could lay more track and make more money from the government subsidies.

Omaha was chosen as the easternmost end of the transcontinental railroad as many eastern train lines ran through there and would make it easier to transfer goods or passengers from the eastern lines to the transcontinental one. Sacramento, on the other hand, was close to San Francisco Bay, the largest hub on the pacific coast at the time.

A transcontinental railroad is defined as a contiguous railroad track that spans the length of a continent. Technically America's transcontinental railroad started well inland and one could argue it isn't a true transcontinental railroad for that reason. However, it has connections to lines that stretch further eastward and the sheer scope of the project at a time when rail travel was still relatively young is demanding of respect.

Now this railroad has become a symbol of westward expansion and an America on the rise. It's been honored in various films, such as How the West was Won and Wild Wild West. The joining of the two halves at Promontory Summit, Utah was even said to have inspired Jules Verne's classic Around the World in Eighty Days. There's no denying the powerful influence something so seemingly mundane has had on our culture and economy.

*Look How Far We've Come*

Since its inception, railroad travel has seen many dips and rises in popularity. For instance, the Great Depression brought about a dramatic reduction in ridership while World War II saw the largest ever ridership, as soldiers were sent off to war from various parts of the country. Its demand has fluctuated with the changing economic and technological climate.

These days, despite the growing presence of cars and planes, the American railroad is still alive and well, offering travel that's generally faster than a car but less expensive than a plane. People still take commuter rails from suburban towns to hubs like Boston or Providence. And they still buy tickets for the Amtrak to travel from Massachusetts to New York or from Denver to San Francisco. What's more, railroad travel continues to have a future as new technological advances promise to keep it relevant and, in many ways, preferable. High speed trains now run at speeds greater than 124 miles per hour. In fact, the Japanese Shinkansen trains are reported to go between 150 and 200 miles per hour. There is no faster way to travel by land.

The continued presence of the railroad in our lives and the undeniably important role it played in the development of our society makes it clear to see why railroad conductors are such a highly romanticized and respected position. As we learn more about these conductors and the vital role they play in the operation of a train, we should remember the rich history from which they stem.

## Chapter Two: The Origins of the Conductor

Maybe you remember seeing the conductor the last time you were on a train. He would have been the man in the suit with the billed hat helping an older gentleman with his luggage or that woman with the collared shirt and big smile collecting your ticket. The conductor is most people's first and only experience with train personnel. Acting in a customer service position is one of many conductors' varied responsibilities. Later in this book, we will address the many hats the conductor wears and what distinguishes this from other roles on the train. However, for now, we're going to take an historical look at the working environment of the railroad, the various jobs performed therein, and the varied challenges they've faced. In doing so, we will learn how this vital job was formed and has evolved into the conductor you know today.

*Early Railroad Workers*

Along with the first railroads in the early to mid 19th century, there came about the first railroad workers. These included engineers, firemen, carmen, brakemen, and conductors. Many of these positions have changed dramatically, having had to adapt to changing needs due to technological advancements. Before we explore the role of the conductor and how it began, let's take a moment to consider some of these other roles and give ourselves a broader understanding of how early railroads functioned.

Railroad engineers still exist today and are also known as engine drivers or train operators. In essence, their job is to drive the train. Given that trains run on tracks, engineers aren't responsible for typical steering. However, they do control the acceleration of the train, stop the train when necessary, and handle it as it runs, tasks which are much harder than they sound.

Railroad firemen used to have the essential job of shoveling coal into the railroad furnace to keep the machine powered. However, that particular need no longer exists on most trains thanks to electric engines. You might think that would put an end to the railroad firemen, but there are still some out there today. They either work on the few steam locomotives still running, or they work on trains with diesel engines where they continue to manage the energy needs of the train.

Carmen tend to the mechanical needs of railroad cars. Although the nature of these cars have changed, they still need regular tune ups and tests to ensure that they are in good condition. Carmen are essentially the maintenance men in the train world.

The aptly named brakemen used to be responsible for literally applying the brakes for individual train cars. This practice has long since been made defunct thanks to automatic braking. But brakeman can still be found in trains today where they perform a variety of tasks including the throwing of track switches and making announcements to passengers. The role has essentially become a catch-all position that has very little to do with actual braking. The name has been preserved largely due to a sense of tradition.

This brings us to the train conductor. Around the beginning of the 19th century, when this position was first conceived of, it was considered the most prestigious railroad job and was consequently the highest paid. The train conductor was the captain of the train. He would dress impeccably, greet passengers, collect fares, make announcements, and generally made sure that everything went according to schedule.

*Fresh Off the Boat*

Maritime tradition places a great deal of value on the role of the captain. The captain is more than simply the ship's "manager."

He or she is the ship's leader. This distinction affords the role a high level of respect. It is up to the captain to ensure the productivity and well-being of the crew. It is up to the captain to guarantee the passengers have a safe, comfortable, and punctual journey. He may not literally steer the ship, but he is responsible for it running smoothly.

It's no accident that we referred to the conductor as the captain of the train. The role of the conductor in fact comes from maritime tradition. Before trains gained traction worldwide, boats were among the primary forms of mass transit. The growing popularity of the railroad threatened to take more than just the passengers from ferries and the like. They also siphoned away many workers who enjoyed similar roles on land as they did at sea. Several of these workers included small time captains who were among the first railroad conductors.

If you've ever heard the popular adage that a captain must go down with his ship, then you must understand the gravity attributed to the role conductors played on early railroads. As leaders, they were charged with putting the safety and comfort of others first, a job they took very seriously.

*Working Conditions*

We will address modern working conditions and how they evolved later in this book. However, given the significant role that railroad unions played in the lives of conductors, it bears mentioning here as well. Conditions were fairly poor for railroad workers in the 19th century. These included inhumanly long hours, low wages, and a high chance of injury. Every railroad job faced its own challenges and potential hazards.

Despite having somewhat higher pay than most other railroad workers and a comparably less dangerous role, railroad engineers were the first to unionize. In 1863 the Brotherhood of Locomotive Engineers was formed. The union played a key role in preventing wages from falling further than they already had in the wake of the Civil War. However, their greatest legacy was likely paving the way for other railroad unions to form.

Brakeman had one of the most dangerous rail jobs in the business. Manually applying the brakes on a train car meant climbing on top of the car while it was in motion and lowering the brakes. It goes without saying how hazardous this could be. The Brotherhood of Railway Brakeman (BRB) was founded in 1883 but was powerless to change this particular aspect of the role as climbing on top of the train was an essential part of the job. However, they were able to fight for higher wages and better hours. They also offered death coverage to help the families of workers in the event of an all too likely accident.

Firemen had possibly one of the most strenuous jobs in the railroad industry. Moving coal demanded significant strength. What's more, the job was an unhealthy one. Firemen were constantly breathing fumes from the coal fires that powered the train. As if this all weren't enough, there was the constant looming threat of an explosion if there wasn't enough water in the boiler. The Brotherhood of Locomotive Firemen was founded in 1873. It eventually expanded and became the Brotherhood of Locomotive Firemen and Enginemen in 1906 as many of its members went on to become train engineers. Like the brakemen union, the B of LF offered death and injury benefits, knowing that they could do little to prevent the hazardous conditions but could at least offer compensation if something were to go wrong.

Although the job of the carman was not as dangerous as that of the fireman or brakeman, it still suffered from long hours and low wages. Inspired by the unions that had come before it, the Brotherhood of Railway Carmen, founded in 1890, sought to protect the interests of its members and raise their prestige within the railroad industry.

Conductors too faced less than ideal working conditions despite the level of respect afforded the position. The Order of Railway Conductors of America (ORC) was founded in 1868, shortly after the Brotherhood of Locomotive Engineers. Although their jobs were not quite as strenuous or as dangerous as many others, they still faced long hours, arduous journeys, and potential hazards as railroad safety standards were still being developed. In addition to representing their interests and providing them with insurance, the ORC worked to promote the conductor as a prestigious occupation and heighten the standards to which they held themselves.

*Living Tradition*

While they still play a key role, particularly on passenger trains, the conductor has survived in America to this day as an occupation largely thanks to railroad unions and the desire to preserve a staple of the American landscape. We romanticize the role in stories and artwork, expecting our conductors to dress in a certain style because, for us, what they do is just as valuable as what they've come to represent.

# Chapter Three: Types of Conductors

When we are talking about "types" of Conductors, we are really talking about types of trains upon which conductors might work. The job may vary slightly to accommodate the particular needs of that type of train, but the essence of the job is largely the same. Conductors are there to manage the train, whether that train happens to be a small trolley in the middle of a metropolitan city or a large Amtrak toting travelers between Boston and New York City. In this chapter, we'll be focusing on the two main categories of railroad train: passenger and freight. There may be some additional variation between different types of passenger train jobs or freight train jobs, however, the primary purpose and functions of each of these types is largely consistent.

*Freight Conductors*

A freight train is a train that transports goods as opposed to passengers. Freight trains in the United States carry everything from chemicals to food. Walk into your local supermarket and investigate the produce section. The fruits and vegetables you see there, unless native to your state, have likely been transported by train from other parts of the country. Freight trains have been in operation since the beginning of the 19<sup>th</sup> century, making the role of the freight train conductor just as old and established as that of the passenger train conductor.

Today, as in the past, the freight train conductor is the authority on the train. This is true no matter how old the individual is or how long they've worked in the industry compared to their colleagues. The conductor is responsible for the contents of the train and for making sure it arrives at its scheduled location on time.

Freight train conductors must know what their train is carrying at all times. This includes understanding any relevant safety precautions and filling out any relevant paperwork. As the train makes its stops, this conductor must know what needs to be unloaded and where. In short, it's a more management oriented position and is far removed from the stately conductor you see portrayed in Shining Time Station.

Freight train conductors historically got their start as brakemen. As we've already established, brakemen have very different roles these days than in the past. They no longer climb on top of train cars to manually drop the brakes. Instead, they are treated as a general entry level train position. Brakemen inspect brakes, couple and uncouple cars, and occasionally throw manual track switches. From here, workers can develop a familiarity with the train and its day to day operations while working their way up. These days, brakemen are rarely employed on freight trains and the freight train conductor is likely to perform these duties him or herself.

In times past, freight train conductors would situate themselves in the caboose of the train. Although nearly all modern trains no longer have these, you're probably familiar with this classic train car. It continues to be portrayed in children's stories and toys and can still be seen today on the few steam locomotives still in operation.

In the caboose, freight train conductors managed their paperwork and studied their time tables, generally staying on top of their schedule. Additionally, they might switch cargo with other trains, requiring even more paperwork to keep track of what switches were made and when. Any visual inspections of the train happened before it departed. Once en-route, the conductor would remain in the caboose, unable to easily traverse the train as cargo cars, unlike passenger ones, did not lend themselves to much movement. In fact, so isolated was the conductor that before radios became a common practice on trains, the conductor and engineer would be forced to communicate through a series of whistle signals and this might have been the only human interaction available to the conductor for hours.

After freight train safety laws were lightened in the 1980s, most trains lost their cabooses. It was no longer a requirement that they have them or that they staff full crews. Higher tech trains required fewer people to manage them and fewer people meant less need for a train car whose sole purpose was to house train crew in the back of the train. What's more, it is no longer necessary for freight train conductors to monitor the back of the train given the invention of a black box that does this task for them. These days, the conductor spends his time riding in the front of the train with the engineer. It may be close quarters, but it beats being alone.

If you're interested in becoming a freight train conductor, take a look at the chapter on conductor responsibilities. It will outline the duties of this type of conductor more thoroughly. Know that this can be a physically tasking job due to both the long journeys involved in freight travel and the physical strength required to perform some of the tasks.

*Passenger Conductors*

A passenger train is a train that transports people from one location to another. Like boats and airplanes with similar purposes, they seek to provide customers with an easy and comfortable experience. Despite seeing a great deal of competition from the aforementioned modes of transportation, passenger trains are alive and well across the world. In the United States, the long distance passenger train industry is dominated by Amtrak, a government funded organization whose lines stretch to every corner of the United States and hit most major cities.

Passenger train conductors are similar to freight train conductors in that they are the authority on their trains. Also, like freight train conductors, they are tasked with efficiency, punctuality, and safety. They too complete paperwork en-route to their destination. They too run safety inspections and remain alert to signals and switch positions that could affect the safe movement of the train.

The primary difference between passenger train conductors and freight train conductors involves customer service. Passenger train conductors take tickets, open and close doors for passengers, and make announcements. They are many people's first and last experience with train personnel and represent their company with distinction.

Depending on the size of the passenger train, it may also employ one or two assistant conductors. The primary conductor delegates his or her responsibilities to these assistants. Their job function is, in essence, to share a complex and demanding workload. They can also help on long journeys by taking the proverbial helm when the primary conductor is on a break.

Passenger train conductors may also receive assistance from other service personnel. Service attendants may distribute refreshments or assist with passenger queries. Porters often assist with the loading and unloading of passenger luggage. The number of service personnel and extent of their duties may depend heavily on the size of the train and length of the journey.

These conductors are not limited to Amtrak and long cross country trips. Passenger train conductors may also be found on transport as small and simple as a city trolley. Some subway systems, like the New York City Subway, still employ conductors as well, despite the fact that many of the conductors' duties on these smaller trains are either automated or performed by the driver.

If you're interested in becoming a passenger train conductor, take a look at the chapter on conductor responsibilities. This job does not have the same physical requirements as that of a freight train conductor, however, it is still very demanding and involves working with a wide variety of people. One might argue that it takes the most patience.

*A Quick Comparison*

We've already covered the main difference between these two conductor professions, however, if you look closely at these positions, you'll notice that there are some key differences worth considering beyond customer service.

Freight train conductors:

- Generally travel longer distances than passenger train conductors.

- Tend to work in greater isolation.

- Have more physical demands placed on them.

- Receive little assistance from other personnel (typically freight trains only employ one conductor and one engineer these days).

Passenger train conductors:

- Travel long and short distances depending on the type of passenger train.

- Work with people (even if they don't have assisting personnel, they still have passengers to attend to).

- Have fewer physical demands placed on them.

- May receive assistance from other personnel depending on the type of train and length of the journey.

These jobs may have similar names. But if you scratch the surface, you'll easily uncover very different positions for very different temperaments. Take a moment to consider what it would be like to travel across the United States in a freight train with only one other person. Imagine having to stay alert that whole time with no one else to delegate to or even talk to half the time. It's an important job but not one for those easily made lonely.

Now consider the alternative. It's your responsibility to make sure the trains run on time. However, herding a hundred passengers feels a lot like herding cats. They're confused about where to go, where to sit, or who to pay. They have questions, concerns, and complaints. You start fantasizing about that career you passed up in freight train conducting, thinking that it would be so much easier to just latch on a batch of cargo that can't talk than to try and be on time with this circus.

Don't get me wrong. If you're interested in a career as a conductor, you're barking up a very noble tree. However, it never hurts to think about the real challenges each of these professions is most likely to face when deciding which conductor path you want to walk down.

## Chapter Four: Responsibilities

When we imagine train conductors, we imagine men in smart suits with pocket watches blowing whistles as they usher passengers onto cars. They shout "all aboard" before disappearing onto the train themselves and sliding the door closed behind them. The role seems charming and, in a way, even laid back. When conductors walk the train platform, they stroll. When they check their watch, it's never with a hurried expression. And they always smile at the passengers.

While being a conductor could be a relaxed and pleasant job for the right person, let's take a moment to disabuse ourselves of the notion that this is in any way an easy job. The responsibilities of the conductor are varied and, at times, complex. They require vigilance, patience, and a meticulous attention to detail. Rather than list all of the responsibilities in one large chunk, we've divided them up into four main categories. Remember from the last chapter that not every conductor will do all of the responsibilities listed here. Duties may vary depending on the type of train involved or even that company's particular best practices.

*Customer Service*

This is primarily the realm of the passenger train conductor. That's not to say that these conductors don't have a wide variety of other responsibilities. But freight conductors, who don't deal with human cargo, don't have to concern themselves with any of the following.

- Ushering passengers on and off the train.
- Taking tickets or issuing them to passengers already on the train.
- Making announcements to passengers.
- Answering passenger questions related to travel.
- Handling unexpected delays due to passenger issues (passengers may fall ill or have an accident).
- Keeping an eye out for suspicious activity.

Think of the conductor as a host. Although the conductor has many more responsibilities behind the scenes, he or she is also there to make the passengers feel comfortable and welcome. Train travel is an aesthetic experience as much as it is a functional one.

*Crew Management*

As modern freight trains generally only have a conductor and an engineer, there is not much crew for the conductor to manage. Occasionally, a brakeman may be available to assist, but typically the conductor fulfills both his responsibilities and those that would normally have been delegated to a brakeman, including some repairs and operating train switches. The following duties are more relevant for passenger conductors.

- Delegating responsibilities to assistant conductors.

- Making announcements to crew.

- Delegating responsibilities to other crew members (such as instructing them to make repairs or operate switches).

Freight train conductors do manage on site crews like the switch engine crew when trains are in the yard. They also coordinate the loading and unloading of cargo with the help of onsite crews. One could say the role is as much about diplomacy and cooperation as it is about delegation and management.

*Train Management*

When we talk about train management, we are talking about overseeing the operation of the train. This means knowing what's on it, where it's going, and when. It also means that the train is operating according to industry standards. The following duties are relevant to both passenger and freight train conductors.

- Making sure the train sticks to its schedule.

- Coordinating the train's movement with the dispatcher and engineer.

- Ensuring that the proper cargo is picked up or dropped off that the proper locations.

- Knowing what cargo is on the train at all times.

- Checking that the crew and train follow relevant safety standards.

- Corresponding with other conductors and stations while en-route.

- Paying attention to signals, such as switch positions, weather, and track obstructions.

Some of this may sound straight-forward, but much of it is easier said than done. After all, sticking to a schedule can be tricky when you must wait for passengers to board or exit a train, passengers who are all intent to move at their own pace. And freight trains don't have it any easier. Moving and checking cars all take time.

Ensuring that the train is running on time and checking for signals is a matter of safety as much as it's a matter of convenience. If one train starts running behind, it can affect the schedules of other trains that share its track. Wayside signals and constant communication between the conductor, engineer, and dispatcher ensure that all trains are aware of any delays or schedule adjustments. Not paying attention to these changes could result in a train collision, which in turn could be fatal.

*Paperwork*

If you're interested in becoming a conductor in order to escape your dull and sedentary office job, allow us to disappoint you. True, the job of a conductor is more active and you won't be spending your hours constantly behind a computer. But if you thought you were going to escape paperwork, think again. Conductors, like most other professions, have their fair share of paperwork to complete while on their journey. Listed below are the most common types. However, each railroad company operates with its own set of expectations and the exact extent and nature of the paperwork could vary.

- Writing reports detailing any incidents or delays.

- Reviewing waybills (documents detailing the contents of a train).

- Reviewing consignment notes (documents containing shipping information for specific cargo).

- Keeping logs detailing all cargo and stops.

- Reviewing work orders.

- Reviewing bulletins which may contain pertinent information for the journey.

- Reviewing switch lists that keep track of the changing of cars on the train.

Although it may not sound the most exciting, paperwork is the conductor's best friend. Conductors must be organized in order to keep track of shifting schedules and cargo. Waybills, consignment notes, work orders, and logs help conductors manage all of this information. Careful documentation can also help conductors in the event there is a complaint. Keeping thorough track of the train's movement can protect conductors when others try to make unfounded claims.

*Manual Tasks*

Conducting a train is a manual pursuit in general. Therefore, it seems a bit odd to have a separate list for manual tasks. However, it bears repeating that there is a significant physical aspect to this job (especially for freight train conductors). While some of these may seem easy, others require no small degree of physical strength.

- Controlling the train's movement when it is running in reverse.

- Coupling and uncoupling cars (attaching and detaching them).

- Switching (also known as Shunting), which puts cars in position to be coupled and uncoupled.

- Checking that the train cars are clean and undamaged prior to the start of a journey.

- Making sure equipment (such as doors) are working properly prior to the start of a journey.

- Assisting crew in making repairs whenever able or necessary.

- Assisting engineer in checking brakes.

While some of the aforementioned tasks do require a degree of physical strength, you don't need to be Hulk Hogan to become a train conductor. That being said, if you're not one for physical activity and you find any amount of reasonable heavy lifting to be too much, this may not be the career path for you. Typically, train crew are expected to be able to lift at least fifty pounds on a regular basis. They may be asked to lift closer to eighty or ninety pounds on the rare occasion. This is true of conductors as well; their position of authority does not make them immune to hard work in the train yard.

*Jack of All Trades*

As you can see, the job of the train conductor, whether freight or passenger, is a complicated one. What's more, it involves everything from people skills for managing passengers and crew to mechanical skills for assessing the safety of the train and assisting in repairs.

While the following are not set in stone requirements for the job, they certainly couldn't hurt.

- Attention to detail.
  - This is about more than just paperwork. A strict attention to detail will help a conductor better note any safety issues on the train when running inspections. What's more, conductors who are detail oriented will have an easier time noticing suspicious activity.
- A good memory for details.
  - This is particularly relevant for passenger conductors who may find themselves inundated with questions about arrival times, routes, and policies. If a conductor has to constantly consult the books every time a passenger asks a question, it will appear unprofessional and slow things down.
- Ability to multitask.
  - A conductor's attention is constantly being pulled in multiple directions. This is especially true of

passenger train conductors who need to manage passenger complaints and requests in and around managing the train, doing their paperwork, and more. However, freight train conductors can also find themselves pulled in multiple directions. Even as they are looking over their paperwork, they need to be alert for communications regarding schedule changes or wayside signals that could affect the trip.

- Patience
  - Anyone who has ever worked a position in customer service would say that this goes without saying. After all, travelers can be an anxious bunch and that tension can easily turn into lashing out. Passenger train conductors must keep a cool head to be able to diffuse situations and maintain a professional demeanor. Freight train conductors also need patience, just a slightly different kind. While remaining vigilant during long journeys, it is easy to allow your mind to wander.
- Problem solving skills

- o Anyone can make a decision or have a knee jerk reaction. Conductors, however, need to make smart and reasoned decisions (sometimes under pressure). This is not a job for people who like mindless, automated work. Nor is it a job for people accustomed to following directions. Conductors must be able to think critically and trust in their ability to make the right choices.

- Organization skills
  - o Train conductors have to be on top of a variety of things. This means keeping track of tickets, money, and more. It also means keeping careful records regarding schedules and cargo. This may seem redundant, as we've already covered "attention to detail" and "a good memory for details." But it's worth mentioning this skill on its own. After all, it's not enough to just pay attention. Conductors need to be meticulous about keeping track of things and they need to know where their paperwork is at all times so as to access it with little trouble.

- Clerical skills

  o This overlaps with "organizational skills" somewhat. However, Clerical skills are about more than just filing out your paperwork in a logical manner. Clerical skills include things like data entry, working phones, and being competent with basic computer programs. In short, a conductor needs to understand the tools related to the more administrative aspects of his job and how to use them most effectively. This includes understanding the paperwork and filling it out correctly.

- Mechanical skills

  o Nobody is asking a prospective conductor to be a mechanical engineer. However, it would serve any prospective conductor well to have some basic knowledge when it comes to mechanical systems. It also doesn't hurt to have an aptitude for mechanics, as that will help anyone looking to learn on the go. As most conductors work their way up from a lesser position on the train crew, they generally already have some working knowledge of train systems.

If you are seeking employment as a conductor and feel overwhelmed by this list of tasks and qualifications, fret not. As with any position, skilled or otherwise, there is a degree of "on the job" training. This training is extensive and can sometimes last for months. It includes instruction on day to day tasks, train operating procedures, safety regulations, and more. Trust that train companies do not want to throw their conductors in the deep end. There's no place for a sink or swim policy when the safety of people and equipment is at stake.

## Chapter Five: Modern Working Conditions

We've already spent a great deal of time talking about what train conductors do. So it's not difficult to imagine what their working conditions may be like. However, given the relative risks involved in such a career, it's important to outline exactly what those conditions are. To start, we'll be taking a look at a day on the life of both a freight train conductor and a passenger train conductor. In this way, we'll get a better picture of what it might be like to have one of these roles.

Keep in mind that when we talk about the day in the life of a conductor, we are using the word "day" very loosely. Trains, whether passenger or freight, may travel at all times during the day or night. Therefore, when a conductor starts his or her "day," it might be eight in the morning or it might be eleven at night. What's more, the day doesn't end exactly eight hours later the way you might expect a typical office job to end. Depending on the line and the length of the journey, conductors could be away from home for more than twenty-four hours.

Also keep in mind that every conductor job can be subtly different depending on the company that conductor works for. Some jobs may promise more regular hours or more reasonable days right off the bat, particularly if the distances traveled are short. However, the following description should give you a general idea of what you're likely to expect as a conductor and most of the activities described will apply to most conductor positions.

*A Day in the Life of a Freight Train Conductor*

A freight train conductor gets a call and is told when his next trip or "tour of duty" is. When he arrives, he first reviews all of the paperwork for that trip, making sure that everything is in order, including that the train is authorized to use the track. Once he's finished checking and double checking the paperwork, he briefs the crew on the details of that journey and any relevant safety measures. Finally, he conducts a physical inspection of the train.

At this point, things can vary somewhat. The next step is to ensure that the train is loaded with the proper cargo. This could involve waiting for another train, traveling to a specific location to acquire the cargo, or simply shifting train cars to make sure the correct cars are attached to the train. Shifting could involve walking on ballast (the material, usually stones, upon which the track is laid), riding end ladders along cars that are being moved, or coupling and uncoupling train cars.

Once the train is ready to go, the conductor spends his time keeping an eye out for signals and obstructions, reviewing paperwork, and staying in constant communication with the engineer, dispatcher, and other conductors. This doesn't have to be as tedious as it might sound. Freight train conductors who love their jobs appreciate the freedom of traveling through open country. There are, in fact, parts of the country that can only be seen by rail. And for many, the fresh air combined with minimal supervision can make this a very appealing position.

Once the cargo is delivered, the conductor will likely spend the night in a local hotel or motel before traveling back home. He may already have a train upon which he will make the trip or he might await a call to confirm his assignment on a train. There is an element of uncertainty involved and freight train conductors must be flexible and patient when it comes to these assignments. This can be hard on people with families or other responsibilities back home.

When conductors get "rides home," they may be the conductors on duty for the trains to which they are assigned or they may be "deadheading." A deadhead is a crew member who is just along for the ride. During this time, the deadhead conductor is still considered "on duty" although he or she is not responsible for the conducting duties.

Someone who is at home in this type of position might be carefree but responsible. They may enjoy the freedom of traveling through the country without having a supervisor looking over their shoulder while remaining vigilant and authoritative whenever necessary.

*A Day in the Life of a Passenger Train Conductor*

Passenger train conductors may also start their days very early or very late depending on the train schedule. When their day begins, their initial duties are not all that different from those of a freight train conductor. He or she will review the paperwork for the trip, including timetables and any other instructions from the dispatchers. The conductor then reviews this information with the train crew, making sure everyone is up to speed on the scheduled movements of the train.

Most passenger conductor roles will also require you to conduct a physical inspection of the train, checking for safety violations and generally making sure it is ready to receive passengers. Some shifting may be required as passenger conductors assist in the coupling and uncoupling of train cars. During this time, the conductor will also assist in checking brakes and making minor repairs.

Once the train is ready to go, the conductor manages its departure as it goes to pick up its first load of passengers. As the train stops to let the passengers on, the conductor opens the doors and ushers them onto the train. He or she might also assist them with luggage or direct other crew members in assisting them. If relevant, he or she also ensures that passengers locate the correct train car. Finally, the conductor closes the door and instructs the engineer to depart.

While en-route, the conductor takes tickets, answers questions, and handles passenger emergencies. The conductor also makes announcements relevant to train conditions, schedule delays, and upcoming stations. When not handling the passengers, the conductor is reviewing timetables and keeping an eye on the track.

Passenger questions and complaints can be varied. A passenger conductor should expect to be met with potentially antagonistic passengers. He or she should also expect to answer the same questions repeatedly as passengers arrive and leave. Finally, conductors may be forced to eject passengers from the train who haven't paid their fare.

This process repeats as necessary until the train makes its final stop. Passenger train conductors usually find themselves back home at the end of the day (albeit usually after a long shift). However, some might find themselves spending the night in a hotel or motel as they wait to work or deadhead their way home. This entirely depends on the length of the route.

Someone who is at home in this type of position might be patient but authoritative, willing to listen to passenger complaints without giving in to unreasonable requests or allowing passengers to ride for free. They may be friendly, poised, and eager to serve.

*Physical Risk*

Train conductors, whether for freight or passenger trains, may not find their job as dangerous as that of a police officer or fire fighter, but make no mistake. When working with heavy machinery every day and traveling at high speeds, you face no small amount of risk. While this should in no way discourage interested people from the profession, it's worth understanding what you might be facing if you become a conductor.

Most of the risky activity associated with train conducting involves shifting (called shunting in Europe). This is when the conductor aids yard crew in moving and attaching cars.

During this time, conductors may find themselves walking on ballast. As was previously mentioned, ballast refers to the bedding upon which train tracks would be laid. At first glance, ballast doesn't look all that dangerous. In fact, it just looks like small stones. However, it can be hard on the legs to walk on for extended periods of time. What's more, ballast can sometimes be large, making it hazardous to walk on, especially in poor weather.

Shifting can also involve riding ladders on the sides of train cars. Although it is no longer policy to have conductors get on and off these cars while they are still moving, this can still be a risky activity. It may not look it, but the ladders are high off the ground and a fall from one could cause injuries.

Not all safety concerns have to do with shifting trains. Many trains transport hazardous or highly flammable materials, such as oil. Proper adherence to safety standards is particularly important in these cases as a collision or derailment could result in explosions, fires, and massive loss of life. Unfortunately, some railroad companies have taken to cutting corners, so to speak, in an effort to meet high demand. As a result, some trains may be too long or use outdated equipment. It is important for conductors and engineers to understand the dangers posed by their particular train and take all possible precautions.

Fortunately, passenger train conductors do not have to worry about hazardous materials. That being said, transporting people can come with its own risks. Typically, passengers will be reasonable and adhere to the train's rules and regulations. When a passenger becomes frustrated or uncooperative, the biggest concern for the conductor is just the headache he or she is likely to have at the end of the day. However, it is possible for a passenger to become belligerent, and a prepared conductor should know how to diffuse a potentially explosive situation. He or she should also be familiar with the proper protocol for handling aggressive passengers, unattended bags, or any other concerns related to passenger safety.

*Compensation*

Train conductors tend to make a decent salary right out of the gate. This is partially due to the potential dangers of the job and partially due to the long and odd hours they may be working. Although these factors are consistent across state lines, salaries may vary depending on location and the train company in question. We will be talking about conductor compensation in terms of salary. However, conductor positions are "non-exempt." In other words, they are hourly positions and not exempt from overtime. Therefore, when we are talking about annual salaries, we are really referring to the effective salary a conductor would make in a year based on his or her hourly earnings. What that conductor actually makes in a year may vary depending on the number of hours worked.

Starting salaries can be harder to pin down than average ones. What a company offers for a starting salary depends on a wide variety of factors. Different companies offer different pay scales. These scales may fluctuate according to location and the cost of living in that location. Finally, salaries may depend on experience. However, a quick glance at available train conductor jobs reveals a low end of roughly $45,000 to $55,000 for new recruits (that's between $21 and $27 an hour give or take).

According to Glass Door, railroad conductor salaries range from about $45,000 to $88,000 nationally, averaging to about $66,238. However, Glass Door also presents salary ranges and averages by company, drawing from reports of specific salaries to generate that data. With 63 salaries reporting, Union Pacific pays it conductors between $40,000 and $100,000 annually, averaging to about $64,774. Also with 67 salaries reported, BNSF Railway pays its conductors between $47,000 and $105,000, averaging about $74,426. While Glass Door did report on the average salaries for other rail companies, many of the sample sizes were too small to be reliable.

According to the United States Bureau of Labor Statistics as of May 2015, train conductors and yardmasters earned salaries ranging from $38,450 to $77,940 (or $18.49 to $37.47 an hour), averaging to about $56,760. The US Bureau of Labor Statistics also tells us that the states with the highest average wage for railroad conductors and yardmasters include New Mexico, Arizona, Mississippi, Minnesota, Wisconsin, Indiana, Kentucky, South

Carolina, and New York. The states with the lowest average wage include Colorado, Kansas, Ohio, Pennsylvania, West Virginia, North Carolina, Florida, Massachusetts, and New Hampshire.

Although salary can be somewhat of a moving target, it's clear that people in this line of work can expect a reasonably comfortable living. What's more, conductors (as with most rail workers) enjoy good benefits. These include health insurance, disability insurance, family assistance programs, and pensions. Like salary, benefits may vary and depend largely on the unions.

Train conducting can be a terrific and lucrative line of work for anyone with or without a college degree. However, remember that these benefits and pay scales do not come without a price. Before they've thoroughly established themselves, conductors often find themselves working weekends, holidays, and more.

# Chapter Six: A Look at Railroad Disasters

Since we're on the topic of modern working conditions, it behooves us to take a moment to talk about exactly how dangerous railroad work can sometimes be. This is not the kind of job where a workplace accident results in a bad back or a broken leg, although both those things can happen. Railroad accidents can be deadly for crew members, passengers, and bystanders. What's more, hazardous materials can have devastating environmental impacts when spilled. While train accidents are rare, they do happen and they are very serious. It's important to understand how much damage trains can do, especially if you're considering going to work for a railroad company.

If you look at the list of American railroad accidents on Wikipedia, you might feel a bit heartened. Sure, there are several dozen there, but they date back to the mid-19th century. Surely other industries have seen more accidents in a year than these railroads have seen in their entire existence. First, these are not listing every broken bone, only times when the train crashed. It is possible to injure yourself on the railroad when doing nothing more than walking on ballast. Second, you'll notice that the accidents in the first sixteen years of the twenty-first century total nearly as much as the entire last century. In other words, these accidents are becoming increasingly more common and are, therefore, of greater concern.

*Runaway Hazard*

If a derailed train weren't bad enough, the damage done by these accidents is sometimes increased thanks to the hazardous materials on board. Back in the early days of the railroad, trains might have transported coal or iron. Sure, these materials were heavy and a train carrying coal would certainly do some damage if it tipped over, but there's a far cry between heavy and explosive. These days, freight trains might carry such flammable or explosive materials as oil and various petrol chemicals used in manufacturing. Even fertilizer can be extremely dangerous when involved in a train accident. This makes freight train accidents exceedingly dangerous not only for those on board, but also for people in the surrounding area and the environment. When a train spills oil into the surrounding area, it is, in effect, an oil spill. In June of 2016, a Union Pacific train crashed in Oregon, spilling about 42,000 gallons of crude oil into the surrounding country and, subsequently, a nearby river. You need only look up the dozens of news reports and documentaries featuring oil drenched birds and fish to appreciate what this means. Wildlife along the river grew sick and died while the river itself was rendered uninhabitable. Of course, oil doesn't need to find a river to do substantial damage. The breached train cars caught fire and oil leaked into the top soil surrounding the accident, killing plants and hindering further growth.

This is hardly an isolated incident. In November of 2015, a train car carrying crude oil spilled when dozens of such cars derailed in Watertown, Wisconsin. While one car resulted in significantly less oil being spilled than in the more recent incident, it was still enough to warrant the evacuation of a nearby town. Admittedly these accidents are far less likely to result in human deaths than passenger train accidents. But they are common, with at least seven such crashes having occurred in North America during 2015 alone.

*Lac-Megantic*

One of the deadliest North American train accidents in recent history and the deadliest in Canada to not involve a passenger train took place on July 6th, 2013 near the town of Lac-Megantic in the eastern townships of Quebec. The train belonged to the Montreal

Main and Atlantic Railway (MMA) and was manned by a single engineer.

The engineer parked the train uphill from the town of Lac-Megantic before turning in for the night at a motel in town. According to his statement, he had applied the main brakes (air brakes) and hand brakes. However, there was some debate after the accident over the validity of this claim and the engineer was later accused of having applied on air brakes and not even enough of those. (The minimum required number of air brakes according to MMA policy was nine and the engineer was accused of having only used seven.) In either case, he stopped the train, leaving the engine running to keep pressure on the air brakes. If the locomotive were to be shut down, the air brakes would no longer be effective.

That is how it came to be that a train carrying crude oil was left running and unattended uphill from a small Quebec town in the middle of the night. Of course, this sounds bad but, short of the controversy regarding the brakes, this was all well within MMA regulations at the time.

Somehow or other, the train went into "distress." In other words, the lead locomotive, the one left running to ensure that the air brakes continued to function, was starting to smoke. Something was clearly wrong and numerous passersby reported having seen the smoke and been concerned about the state of the machine. How the engine went into distress is still a mystery. MMA claims that the train might have been tampered with when left unattended and unlocked. However, these claims have never been substantiated. It was also reported that the engineer was made aware of the state of the train and, after communicating with dispatch, was told to leave it until the morning.

All that really matters is that the engine shut down sometime during the night and the brakes went with it. The train picked up speed as it rolled downhill along the tracks. As it neared the downtown of Lac-Megantic, it's speed was so great that the train derailed and collided with the town. Given the volatile nature of the train's contents, many of the cars exploded and half of the downtown was destroyed. Later, forty-two people were confirmed dead and another five were missing and presumed dead.

The engineer faced charges of criminal negligence under the allegation that he had not set the appropriate number and type of brakes. However, MMA also came under fire for lax policies that aided in bringing about the disaster. They've since required that all trains containing hazardous materials be supervised at all times (day or night). They've also made is mandatory for two train personnel to be working a route instead of just one, requiring a conductor to accompany each engineer.

While we can't know exactly what happened that night, we can learn something essential from this tragedy. Train personnel have a very important job and one that demands care and attention. Perhaps the engineer, like all of us, was exhausted after a long day. After all, he was alone on that train. When we can't wait to get out of the office, maybe we leave a few emails for tomorrow and rush through the report we need to finish, making some minor errors. So maybe we can sympathize with him. That doesn't change the fact that, on a train, there are no minor errors. If he was lazy about applying the brakes, eager for his day to be done, he's not looking at a slap on the wrist and a little extra time spent crunching numbers the next day.

*Early Train Accidents*

While it is important to appreciate the devastating impact of freight train accidents, there's no denying that the vast majority of deadly accidents involve passenger trains, and this is as much an issue today as it was over a century ago when trains were still new. Despite increasingly more stringent safety precautions and other policy changes, there remains the potential for human error and mechanical failure.

There is some inconsistency regarding the first deadly railroad accident in the United States. If you were to go by, say, the History channel, you would see the 1832 Granite railway accident credited as the first. However, there is some record of accidents occurring in the United States before that time, including the 1831 boiler explosion in South Carolina that killed a crew member. Regardless, the Granite railway accident was likely the first derailment, the culprit behind most deadly train accidents.

On July 25th, 1832, a group of non-railroad personnel were invited to witness the transportation of heavy loads of stone. They, perhaps inadvisably, did this from a vacant train car. When the cable holding their car snapped, the car and everyone in it fell thirty-four feet down a cliff. One of the four visitors died, while the others sustained injuries.

The first head on collision that resulted in the deaths of passengers occurred in 1837 on the Portsmouth and Roanoke railroad. A passenger train collided with a freight train that was hauling lumber near Suffolk, Virginia. Dozens were injured, but the accident was particularly notable because three daughters of an important local family were killed.

Early train accidents increased in fatalities over the years. The Shohola Train Wreck killed sixty people when a train carrying confederate prisoners across country during the Civil War was in a head on collision with another. A few years later, in 1867, the accident that became known as the Angola Horror killed forty-nine people and seven years after that, in 1876, the Ashtabula River Road Disaster killed ninety-two, making it one of the deadlier train accidents in US history. However, despite how obviously tragic these events were, they weren't the worst the US had seen.

*Deadliest Train Accidents*

On November 1, 1918, the Brighton Beach Line Accident, also known as the Malbone Street Wreck, killed ninety-three people instantly and another five, who died later from injuries. The accident occurred in Brooklyn, New York when a subway (or, rather, the 1918 version of a subway) was taking a turn at too high a speed. The turn, meant to be taken at around 10mph, was instead tackled at a whopping 30-40mph. As a result, the wooden train cars were torn apart.

Even deadlier, was the Great Train Wreck of 1918, officially making 1918 the worst year for trains in the United States. Two passenger trains experienced a head on collision near Nashville, Tennessee. Both trains were expected to share the same mile long stretch of track. However, one of the trains was found at fault for not having properly accounted for the other, which had the right of way. As a result, a hundred and one people died and another hundred and seventy-one people were injured.

*It's Not All about Derailments*

Obviously derailments can cause a great deal of damage. However, it's not enough to make sure your train stays on the tracks. Trains can experience any number of other issues that can have expensive or deadly consequences, not the least of which was the boiler explosion we mentioned above, which resulted in the death of a crew member.

On May 4th, 1840, a bridge collapsed under the weight of a train. The train, traveling between Catskill and Cairo New York on the Canajoharie and Catskill Railroad, fell into the water. Thankfully, there was only one fatality. The incident may seem relatively minor but it makes an important point. You can't control everything that might go wrong. Obviously, neither the conductor, nor anyone else on board, could be held responsible for the poor condition of the bridge. It is, therefore, all the more important that the conductor and crew be vigilant and prepared to react as best they can to unexpected situations.

Of course, sometimes accidents that don't involve derailments are still the fault of train crew. The Norwalk Rail Accident of 1853 resulted from human error. The train was due to cross a drawbridge and the conductor failed to check for a signal that would indicate whether the bridge was up or down. As you can probably guess, the train nose-dived into the water below because it planned on bridge that wasn't there. Forty-six passengers died.

*The Moral of the Story*

I know you're probably wondering why we are burrowing so deeply into discussing the more morbid aspects of train travel. In short, it doesn't do to sugarcoat any of the darker aspects of train travel if you're looking into it as a potential profession. True, more often than not, you'll have a clean journey. It's not like conducting a train is some video game where you're dodging obstacles left and right. But knowing what could happen is essential when it comes to truly understanding the gravity of the position you hold. Maybe the conductor in our last example would have been more alert had he known what was at stake. Maybe the Brooklyn driver in the Malbone Street Wreck would not have been so hasty if he knew what could have happened. It's important to understand these disasters and how they came about to foster the proper respect for the position you seek to hold.

## Chapter Seven: Modern Rail Unions

If you are coming from an office job, you may not be that familiar with labor unions, how they function, and what they do to protect the interests of their members. Labor unions are essentially collectives of workers who band together to protect their mutual rights and interests. They tend to come about in industries where workers with particular skills are likely to be taken advantage of or otherwise under-treated. They are also very common in high risk industries where they often appeal for better safety standards.

For example, public school teachers have labor unions. They pay union dues to enjoy the benefits of their union. In other words, a small portion of their paycheck goes straight to the union and this money is used for things like legislation, lobbying, and more. Unions protect teachers from unfair administrative practices, unfounded accusations, and wrongful termination. They also support policies like tenure that help protect teachers.

Railroad unions would operate similarly. Railroad workers are typically expected to pay union dues. The unions then represent the workers in various ways. They might push for higher pay, better safety standards, or legally represent you if the railroad company accuses you of doing something unsafe or illegal. Unions also organize strikes, organizing their members in an agreement to refuse to work until union demands are met.

*The Pros and Cons of Unionization*

Unions can be highly beneficial, but have also come up against much criticism for their practices. Strikes can have damaging effects on industry and economy. Of course, that's largely the point as union want the industry to feel the hurt so that those in charge are more likely to make compromises. However, strikes can also be seen as unnecessarily stubborn or belligerent and many accuse unions of forcing unfair or unrealistic demands on employers.

Some people also criticize unions and employers alike for what some called forced representation. In other words, some employers and unions may strike bargains that require all members working for said employer in a particular role to pay union dues and, thus, be part of the union whether they want representation or not. However, this is largely frowned upon and many states have laws that prevent such agreements from happening.

If you choose to join a railroad union or are subject to one of the aforementioned agreements, get involved. Learn about your union chapter. Meet your representatives. Go to meetings if you can. The union will be taking action on your behalf whether you chose to join or had to join. You might as well figure out what that means and become a positive influence.

*The Evolution of Rail Unions*

When we were discussing the compensation that a prospective conductor might expect, we mentioned in passing that unions have a large role in determining said compensation. Revisit the chapter on the history of the conductor and you'll recall that railroad unions are long established and highly respected institutions. The ORC (Order of Railway Conductors in America) was one of the first railroad unions and sought to promote the interests of conductors. At the time, this meant they wanted "to unite its members; to combine their interests as railway conductors; to elevate their standards as such and their character as men for their mutual improvement and advantage, socially and otherwise..." The ORC sought to have a positive impact on the overall quality life of train conductors and to earn them the same respect and admiration that ship captains and similar occupations.

In the mid 20th century, the ORC merged with other railroad unions to form the United Transportation Union (UTU). The UTU was the largest railroad union in North America and managed the interests of nearly all American railroad workers, including conductors, brakemen, switchmen, engineers, yardmasters, and more. The UTU is still technically around. However, we use the word "was" because as of 2014, it merged with the Sheet Metal Workers' International Association to form The International Association of Sheet, Metal, Air, Rail and Transportation Workers (SMART).

Covering not only the United States but also Canada and Puerto Rico, SMART protects the interest of an even larger collection of rail workers (among other professions). SMART professes to strive for better contracts for its rail employees. This includes higher wages, better benefits, and a lucrative pension. SMART also strives to provide resources to members in need of assistance and to protect those workers from potential hazards associated with their jobs.

*SMART Departments and Leadership*

Given its wide-ranging interests, SMART has a thorough
organizational structure, each with its own goals. Although these
departments are not specific to the transportation portion of SMART,
it's worth taking the time to explore SMART's inner workings in
order to better understand the extent of the union and how it
operates.

- The Communications and Research Department provides
  union leaders with accurate information regarding research,
  policy, and more. Additionally, it supports leaders and
  members when they must speak publicly about issues that
  concern them.

- The Department of Canadian Affairs does pretty much
  exactly what it sounds like. They promote SMART interests
  in Canada.

- The Department of Government and Legislative Affairs
  support members where it matters most by helping to push
  beneficial legislation, supporting politicians whose interests

align with theirs, and teaching members about key political issues that have bearing on their lives.

- The Department of Jurisdiction concerns itself with tradition. It does more than just settle disagreements between jurisdictions; it encourages consistency and integrity within all jurisdictions. In its own words, it seeks to "perpetuate the performance and dignity of trade from generation to generation."

- The Education Department provides training to elected or appointed representatives so that they might better represent union members.

- The Legal Department advises union leadership on all relevant legal matters. It also supports local unions with legal issues and educates members and leadership alike on labor laws.

- The Mechanical and Shipyard Department (also known as the Mechanical and Environmental Department) specifically handles matters related to shipyards and railroads.

- The Organizing Department works on recruiting new union members. It also concerns itself with expanding its market share. In other words, it wants to earn a bigger piece of the pie for its members, getting them consistently higher wages.

The union has three main officers: the General President, the SMART Transportation President, and the General Secretary Treasurer. There are also a plethora of General Executive Council members, General Vice Presidents, and other roles at various levels within the organization.

*International Brotherhood of Teamsters*

Although the United Transportation Union under SMART is the largest railroad union in North America, it is not the only one. The International Brotherhood of Teamsters (IBT) was created in 1903 with the merger of several smaller unions, including The Brotherhood of Locomotive Engineers and Trainmen (BLET), sometimes referred to only as the Brotherhood of Locomotive Engineers.

Although the word "teamster" is commonly used to refer to a truck driver, the IBT represents a wide range of skilled workers in

84

both the US and Canada. In fact, their website brags that they represent everyone from public defenders and brewers to vegetable workers and zookeepers. However, the BLET makes up a large chunk of the represented workers.

*Railroad Workers United*

Even within their respective collectives, the United Transportation Union and the Brotherhood of Locomotive Engineers and Trainmen have occasionally butted heads. So much so, in fact, that Railroad Operating Crafts United (ROCU) was formed. This organization was comprised of non-leadership members of both the UTU and the BLET who wanted to end the divisiveness between the groups.

Now known as Railroad Workers United (RWU), this organization continues to try and bring together railroad workers from different roles. RWU is not itself a union, but seeks to bridge the gap between unions representing different roles in the same field.

*To Unionize or Not to Unionize*

To be honest, you aren't faced with much of a choice. If you plan to work on the railroads, you'll likely be part of the relevant union. But while you may not be looking forward to paying your union dues, you can look forward to some solid representation and a community of people that understand and respect the important role you play.

## Chapter Eight: The Uniform

If you were to look for a train conductor outfit for
Halloween, you'd likely find yourself wearing a formal and old-
fashioned suit, complete with slacks, vest, jacket, pocket watch, and
billed hat. You might think you'd look like a caricature, but you
wouldn't be far from reality. The train conductor uniform is inspired
by a sense of prestige and professionalism. When conductors first
started dressing like Tom Hanks from The Polar Express, they did so
because they wanted to distinguish themselves and elevate the
impression that they left with people.

Although the uniform has evolved somewhat over the years,
for many rail lines, it hasn't changed as much as you'd think. These
days, many conductors retain formal attire partly to maintain the air
of professionalism that they cultivated so long ago and partly out of
a sense of history. This is driven home by the fact that most
conductors who maintain this formal dress and the classic billed hats
work on passenger trains, where they are more in the public eye.

Whether we are talking about conductors in the early 20th
century or conductors today, we are faced with a similar difficulty.
In short, what a conductor wears depends on a variety of factors,

including the rail line for which they work and the type of train they work on. This is truer today than in the past, but the fact remains that rail lines of the past also had their own set of specifications.

Here, we will discuss common threads between past and present train conductor uniforms. We will also examine specific uniforms for the past and present day to give you an example of what exactly a train conductor might have worn.

*Early Conductors*

Very early railroad workers didn't wear uniforms. They wore standard business attire. However, the need for uniforms became apparent when there was some confusion on the train lines. Railroad companies quickly began to recognize the need for a consistent way to distinguish between railroad employees at different levels. Given the aforementioned status of the conductor as the effective "captain" of the train, it was particularly important to distinguish him in an official way.

Railroad companies began issuing strict standards for how railroad employees should dress. Depending on the company you worked for, you might have a double-breasted jacket or a single breasted one. You might have pockets or collars measured at different lengths. However, certain details were largely universal.

Historically, early railroad conductors wore navy blue, this color, among other things, distinguished them from other crew members on the train. They also wore jackets with vests, ties, and collared shirts underneath. Conductor hats were typically billed and sported badges labeling the conductor as such. Finally, they carried the quintessential conductor's pocket watch. At the time, this was a largely practical addition. After all, conductors needed to be punctual and pocket watches were the timepieces of the day. However, this tool soon became a symbol, as conductors continued to carry them long after more practical timepieces were invented.

*Northern Pacific Railway Conductors*

On October 22, 1913, The Northern Pacific Railway Line made official a detailed manual specifying exactly how conductors and other railroad crewmen were to dress. This manual, called the Rules and Specifications Governing the Uniforming of Employees, came complete with photographs and diagrams. According to the manual, all railroad employees were required to wear uniforms and badges in keeping with their position. They were also required to be "neat in appearance." Conductors, in particular, were required to wear matching navy blue slacks, vests, and jackets, underneath which they sported ties and white collared shirts. The jackets had to be single breast with four buttons holes down the front and one on the lapel. The buttons had to be made of black hard rubber. The jackets were required to sport one outside breast pocket, two inside breast pockets, two outside ticket pockets, two inside ticket pockets, and two outside skirt pockets. The pamphlet even dictated the exact dimensions of the pockets and width apart of the buttons. The jackets were meant to be lined in wool but for the sleeves, which were lined in silk. The vest and slacks had similarly stringent specifications, dictating the number of buttons and pockets and the length, stitching, and fabric. The hats required a certain height and were accompanied by a bill. They all had double chords running across the front and just beneath a badge identifying the wearer as the conductor.

## Contemporary Conductors

These days, the formality of the conductor's clothing is largely dictated by the type of train upon which they work. This difference is most pronounced when talking about freight train versus passenger train conductors.

Freight train conductors, given the relative solitude in which they work, dress in a more leisurely manner. As their jobs are more physically demanding, they tend to have looser clothing, easy for getting around in. Freight train conductors also generally wear reflective vests for safety reasons.

Passenger train conductors are generally required to be more formal. After all, they manage passengers and larger crews. It is necessary that they be seen as authority figures and dress is no small part of that. Here, the nature of the dress can vary dramatically depending on the train.

Subways and trolleys that still employ conductors, like the New York City subway, tend to wear simple button down shirts with pants and belts. Occasionally, they might wear ties. This is about as casual as it gets for a passenger train conductor. Larger passenger trains, like Amtrak, require more formality. These conductors likely wear button down shirts and jackets with ties and the classic conductor's hat. You might even see a vest or two out there although it is no longer a fundamental part of the uniform.

*Amtrak Conductors*

To get a better idea of what contemporary passenger conductors might wear, let's take a look at the specifications set by Amtrak for its conductors. Note that, unlike the Northern Pacific Railway, Amtrak doesn't dictate their uniforms down to the smallest stitch. However, this may largely be due to the benefits of mass production.

Amtrak maintains an honored tradition and dresses its conductors in navy blue and white. Like many other train companies in the past and present, Amtrak uses colors and styles to differentiate between roles. Amtrak has its train employees where epaulets for this purpose. An epaulet is an ornamental fixture on the shoulders of an outfit. Former military wear often has them. At Amtrak, the color or pattern on your epaulet is specific to your job. For example, a conductor wears a navy blue epaulet while an assistant conductor wears one that is navy blue with teal stripes.

Amtrak also requires its conductors to wear hats not unlike those of the past. Amtrak conductors wear either Pershing hats or pill-box style hats. The former has a wide crown that fans at the top and a small bill in the front. The later is almost like a miniature top hat, also with a bill in the front. All hats come with badges indicating that the wearer is the conductor of the train, a tradition that has stood the test of time.

*Tradition*

Deconstructing the uniform of the train conductor may seem daunting and unnecessary. But if you're serious about becoming a conductor, it's important to understand and appreciate that which you'll wear given the relationship between the conductor's uniform and the role's rich history. Appreciating the conductor's uniform means appreciating the significant role the conductor has played in American history.

Even people who do not strive to be train conductors can appreciate the tradition. Many railroad enthusiasts collect pieces of conductor paraphernalia. They collect badges, buttons, hats, jackets, and even whole attire. Some collectors will even specialize in particular items, aiming to collect hats or buttons associated with different rail lines or different time periods. Even if collecting isn't your thing, the very fact that some people relish these pieces of history should further drive home how historically significant something as simple as a pocket watch can be.

## Chapter Nine: The Colorful Language of the Railroad

Walk into any greasy spoon in North America, and you're likely to hear some colorful terminology. Maybe two poached eggs on toast goes by "Adam and Eve on a raft" or a cup of black coffee is named "draw one in the dark." At such establishments, grapefruit juice goes by "battery acid," a cracker is a "dog biscuit" and ham and eggs are dubbed "the eternal twins." It's hard to say where many of these colorful phrases originated, but you can still hear them in diners across America.

As it turns out "greasy spoon" belongs to another set of creative phrases, one that makes up the obscure lingo of the North American railroad. That's right, railroad culture goes beyond the dress and drudgery. We've taken some time to compile some of the more common, more interesting and, in some cases, downright bizarre utterances you might here if you take a job on the tracks.

Of course, out emphasis is on the word "might." While some of the phrases are credited as being contemporary lingo, most of them harken back to the early days of the railroad and may not be in common use.

*Terms*

All Black – This essentially means "all clear" and indicates that a visual inspection of a train yielded no obvious mechanical issues. The phrase "all black, well-stacked, going down the clickety-clack" means that the train passed a visual inspection while rolling slowly by the inspector (a roll-by inspection).

Armstrong – This blanket term refers to any equipment that must be operated manually. Usually such tasks require a degree of strength, hence the name. Such equipment could include hand brakes, and certain engines.

Back to the Farm – This term is used to describe any railroad worker who has been laid off, particularly due to slow business.

Bad Order – These rail cars have mechanical defects or other mechanical issues. They are barred from use.

Ball of Fire – This, put simply, is a fast run.

Beanery – This is another term for a "greasy spoon." "Beaneries" are diners frequented by railroad workers and truckers on the road. They may also be used to refer to any cheap restaurant attended on the road. Waitresses at such establishments are sometimes called "beanery queens."

Black diamonds – This antiquated term refers to coal, typically the coal used to power a steam engine. However, it may also have been used to refer to coal being transported by a freight train. "Black diamond" featured in numerous other railroad terms as a reference to coal. For example, train firemen, who historically were responsible for shoveling coal to keep trains running were called "diamond crackers."

Boxcar Tourist – Historically, trains offered a popular mode of transportation for homeless drifters. The term "hobo," although widely applied today to refer to general homeless people, was specifically used to refer to migrant workers who were often homeless. Many of these people would board empty freight boxcars on their way to find work. Railroad workers referred to these people as "boxcar tourists."

Broncos in the Canyon – Some vehicles (like trucks) are able to travel along train tracks. These vehicles have special equipment that helps them use the rails. These vehicles are operated by train employees and are typically there to inspect the tracks ahead of a scheduled train. This is done in cases where the tracks may become suddenly unusable (from potential flash flooding, mudslides, or the like). "Broncos in the canyon" refers to the presence of these vehicles on the tracks.

Carry a White Feather – A train is "carrying a white feather" when you can see a plume of steam coming from the engine. Imagine the classic depiction of a traveling steam engine, with the smoke trailing behind the locomotive and along the length of the train.

Couldn't Pull a Setting Hen Off Her Nest – This derogatory expression refers to old-fashioned and out of date locomotives. This phrase likely suggests that said machine is still in use though it ought not to be.

Cornfield Meet – This is a head-on collision between two trains (or near miss). The event is so called because early railroad collisions would most often occur out in the countryside (near cornfields). "Cornfield meet" may also refer to events where old locomotives were intentionally run into each other.

Curfew – A "curfew" is a scheduled time when no trains of any kind will be operating. This allows maintenance crews to work the tracks, checking wayside signals and other important features.

Dancing on the Carpet – A train worker who is called into the boss's office for disciplinary action of some kind is said to be "dancing on the carpet," presumably because it conjured up images of someone who is anxious and put on the spot.

Double heading – This is when one train is accompanied by two locomotives. This term only applies where each locomotive is manned by its own crew and not when one engineer runs all of the engines while sitting in the lead locomotive. Double heading is usually used with steam or diesel engines and is typically done when one locomotive is not capable of handling the train's weight.

Drag – Also called a "drag freight," these freight trains run slowly because they are carrying particularly heavy loads. Such trains have to be particularly cautious when running on steep hills or sharp curves as the excess weight makes for unpredictable movement.

En Routes – The term "en-route" means roughly "on the way." In Railroad terminology, "en routes" are all trains headed to a particular yard or terminal that are in need of switching. If yard workers anticipate a heavy switching load for the day, they call this "strong en routes."

Getting the Rocking Chair – A train worker who is "getting the rocking chair" is retiring with a pension.

Highball – This is a historic term that refers to any signal that indicates a train should run at full speed. In the past, some wayside signals were just balls hung from poles near tracks. If the ball was at the top of the pole, it was considered "high" and a train engineer knew he could continue at full speed. The term has survived to today despite the highball signal being long retired.

Hotshot – This term is still used today to refer to a train that has been given ultimate priority. Hotshot trains are rarely held up and are often given access to the main line, or track, to get to their destinations.

In the Hole – A train that is "in the hole" is essentially sidelined while another train passes. The hole refers to the side track intended for this purpose.

Meet – This is when one train passes another that is "in the hole" or siding.

On the Ground – A train is "on the ground" if it is derailed.

Piggyback – You might hear this explained as the transportation of truck trailers on a train bed. However, more accurately, it refers to the transport of intermodal containers. These shipping containers carry goods as they move between modes of transportation (ship, truck, train).

Power – The locomotive is sometimes referred to as the "power." Locomotives have numerous slang names, but this is one of the most common.

Shoofly – This colorful term refers to any temporary track that is used to help trains avoid obstacles. This is not the same as when a train is "in the hole" (see above) because those tracks, used by trains to allow others to pass, are not temporary. A "shoofly" may be built when a natural disaster such as flooding or a rockslide makes it impossible to use the normal track.

The Biscuits Hang High – This antiquated term suggests that there isn't much food to go around in a particular location. The image of hanging biscuits that are just out of reach reflects this.

Undesired Emergency – Also called UDEs, an "undesired emergency" occurs when air pressure is released (or escapes) the air brakes. As a result, the emergency train brakes kick in and the train grinds to a halt. The term "undesired" likely indicates that the emergency is inadvertent or false since there is no true reason for the train to be stopping.

Zombie – Sometimes locomotive frames are reused to form new engines. When this happens, the machine is referred to as a "zombie." The term may also be used to describe the act of reusing the locomotive frame.

*Railroad-ese*

The jargon, both historical and contemporary, is referred to collectively as "railroad-ese." It's unclear how such languages develop as a whole, although certain expressions may be explained independently. Regardless, these phrases are part of the tapestry that makes up the American railroad and all of its charm.

## Chapter Ten: Conductors Around the World

In this book, we're primarily focused on train conductors in the United States. However, if you recall from the first chapter, the railroad doesn't begin and end in America. It has its roots in ancient Greece, and Europe has enjoyed a rich railroad tradition for just as long as we have. The role and history of the conductor is just as respected in Europe as it is in America. In fact, Wilbert Awdry, the creator of Thomas the Tank Engine, which in turn led to Shining Time Station, was British. From Thomas and his friends to the famous Hogwarts Express, England too has shown a history for romanticizing trains.

Trains aren't purely a western phenomenon either. Although they may have begun in Europe, many Asian countries enjoyed rail transport as early as the mid-19th century. These days, some of those same countries have pushed the boundaries of rail transportation, surpassing their western neighbors. Although the railroad played a significant role in the history of the United States, it is far bigger than the United States alone.

*Europe*

Although the first locomotive was built in England in 1804, it wasn't until the 1820's that mechanized rail transportation truly developed, starting in England and spreading quickly to the rest of the European continent. Over time, various European countries would develop minor differences in the how their trains and crew functioned. However, the overall effect is largely the same and the tradition of the conductor celebrated throughout.

*India*

The first Indian railway, built in 1853 by the British East India Company, was seen as yet another example of how the British were working to colonize and control India. Therefore, it and the lines built shortly after it, were not built for the benefit of the Indian people. Rather, they were built to allow the British easier access to the country. This practice was continued by the colonial government once it was established.

Railroads in India became increasingly more exploitative as the British used them not only to spread troops around the country, but also to transport Indian goods out of India. What's more, these privately-owned railroads saw a great deal of profit from operating in India, profit that filtered primarily back to Britain.

Despite it's rather inauspicious start, the Indian railroad system has since flourished. It expanded greatly after India acquired its independence. Finally, the rail system could work for the Indian people and not against them. In recent years, India has shown its pride in rail network, celebrating 150 years of existence in 2003. Indian conductors, on both passenger and freight trains, are known as guards. These guards, sometimes called mail guards on passenger trains, are responsible for the entire train, including the route, schedule, and cargo. The guard communicates with the pilot (a.k.a. engineer) using colored flags and two-way radios. Like American conductors, Indian train guards check the train for maintenance issues and sell tickets. These guards even carry around first aid kits, making them first responders in the event of an emergency.

*Japan*

Rail transportation in Japan is said to have begun around the end of the Edo Period. This period in Japan lasted from 1603 to 1868 and was marked by economic growth and a strict policy of isolationism. So strict was this policy, known as Sakoku, that it prevented all native peoples from leaving the country and all foreigners from entering. As you can imagine, this was not an environment conducive to accepting and adapting outside technologies. It's important to understand how world changing the end of isolationism was for the people of Japan.

The impact of the railroad on Japan began subtly. Dutch traders, among the few who were allowed contact with the island, brought word of railroad transportation. Other visitors, including the infamous Matthew Perry who was instrumental in bringing about the end of the Edo Period, brought model trains, sparking an interest in this new development.

In 1872, the first Japanese rail line was completed, connecting Tokyo to Yokohama. Nine years later, Japan saw the foundation of its first privately owned railroad company, the Nippon Railway. Prior to this, all rail lines in Japan had been under the purview of the imperial government. In the decade that followed, several more private companies rose, developing an extensive rail network in the small country.

Around the early 20<sup>th</sup> century, many of these private lines (seventeen to be exact) were nationalized. Under the Railway Nationalization Act, the government took control of these lines. This happened again around the second World War. This time, the government forcibly took control of twenty-two additional companies in the interest of the war effort.

It is clear that just as with the United States, the railroad played a significant role in Japan's recent history. These days, Japan is known for punctual trains with strict rules and regulations. They've also pushed technological boundaries, creating a network of high speed trains known as Shinkansen.

Japanese train conductors hold positions fairly similar to those in the United States. In short, they manage the trains, using hand signals and a process called "pointing and calling" to communicate with other members of the train crew. Pointing and calling is most commonly used by the train drivers, but other crew members will sometimes employ it. It's so commonly used in Japan, that it is sometimes called by its Japanese name, shisa kanko. The practice is born from the philosophy that making large gestures and speaking actions helps keep workers focused and helps prevent errors. When a crewman points and calls, other crew members must react. In short, the behavior engages all of the senses and keeps train employees on their toes. Modern Japanese train conductors also dress in a manner similar to western conductors. They tend to dress formally, in jackets and ties. They also wear the trademark billed caps with insignias on the front. You may also see these conductors with white gloves and whistles.

*China*

The railroad came late to China when compared with the aforementioned eastern nations. The first Chinese railways were built in the late 19th century, decades after they'd been established in India and Japan. This delay has been credited to the Qing government and their skepticism when it came to steam engines and their potential use. Additionally, the Qing dynasty had concerns about the potential impact of railroads on the country's feng shui, a practice that seeks to cultivate harmony between individuals and their environment. This attitude toward modern technologies may have been a contributing factor to China's lack of industrialization, another road block to developing a railroad system.

In the end, trains won out. Technically, the first railway built in China was constructed by a British merchant in 1864 for the benefit of the imperial court. It was promptly dismantled. The first railway to see any actual commercial use in China, was built in 1876. It too was built by the British. If you're wondering what finally changed the Qing dynasty's mind, the truth is that nothing did. This railway was also built by the British and without the consent of the government. It was a political power play, backed by a lesser Chinese official. One might say that the railroad came to China despite the best efforts of a centuries old dynasty in love with tradition.

These days, China has a vast railway network with state of the art high speed and maglev trains. It is one of the main methods of transportation for Chinese citizens, making the role of the passenger conductor a very important one. Chinese passenger trains come in a wide variety of classes and styles. Travelers can stay in first class, second class, business class, or a "high speed sleeper." They can enjoy catered meals or eat at the bar. Even the sleeper trains come with a wide selection of styles to suit every passenger.

Chinese passenger trains typically come equipped with a primary conductor and eight assistant conductors and brakemen. Additionally, a car attendant assists passengers with particular needs. According to travel websites and the experiences of foreigners using the Chinese rail lines, the assistant conductors and brakemen are often found at one end of the dining car. However, Chinese conductors have also been said to be efficient and friendly. Many travelers have credited them with being easy to locate and happy to help. What's more, they and their crew have been known to entertain passengers, particularly on Chinese holidays, with decorations, costumes, and more.

*Common Cause*

Railroad technology came to different regions of the globe for vastly different reasons. However, many of the safety standards and roles on those trains have evolved similarly. And railroad conductors largely enjoy a similar status across cultures.

# Russian
# Conductor

Japan
Railways
Conductor

Australian
Conductor

## Chapter Eleven: Superstitions and Folklore

Many beliefs have sprung up around railroads over the past two centuries. Here are some superstitions railroaders have held over the years. as well as a few of the railroad's more strange (but purportedly true) stories from American folklore and beyond. Superstitions often start with some incident which in itself would have no effect upon analytical minds. Repetitions of such incidents may breed a fear that becomes chronic. A locomotive involved in one or more wrecks is supposed to be jinxed, especially if she has a 13 or a 9 in her number. Enginemen fight shy of her. Many refuse to handle a number 13, or begin a run on Friday the 13th, or have anything to do with 13.

You often hear of accidents that happen to engines numbered 13, but there is no reason to believe that these engines have more than their share of misfortunes. The mishaps to 13s are just publicized more than the others. The most baffling holdup in rail history was staged by thirteen bandits who robbed a Union Pacific express of $60,000 in gold at Big Spring, Neb., in 1887. Not one of the mysterious thirteen was ever identified or captured!

Long ago, superstition decreed that no locomotive should be turned to the left as she rumbled out of the roundhouse onto the turntable. September, the ninth month, was considered the fatal month in railroading. That might have been because the first train fatality occurred on a September day in 1830, when a British Parliament member fell under the wheels of George Stephenson's *Rocket*. A checkup of famous rail disasters shows that more of them happened in summer than at any other season. No wonder sighs of relief went up year after year when September's page was torn off the calendar, indicating that no more summer excursions would be run till the roses bloomed again. The ninth month was believed to be jinxed.

SOME veteran trainmen refuse to sweep out cabooses after dark, lest they bring bad luck. Engineers have a superstition that it is unlucky to step onto the cab with left foot first. There is a reason for this. You mount the cab on the right side with the oil can in your left hand. You swing aboard with your right paw clutching the grab-iron on the cab's right side. If you tried to barge in with your left foot first, you'd have to cross your feet over, or stumble.

When a switchman kicks his foot in a switchfrog, that also is regarded as a bad omen. He invariably goes back and steps over it again, to avoid getting his foot caught in it the next time. When a new man trips over a tie or rail, a seasoned railroader will call him back and make him step high over it "for the good of the service." This is no mere superstition: it is a matter of training the green hind in safety habits.

Seaside folk have a belief that births occur at flood tide and deaths at ebb, on the theory that life comes and goes with the tide. In railroading, we have the strange case of James Luther Shull. Jim was born in the year that the Waynesburg & Washington narrow-gage line was projected. When he grew up, he worked for the W&W, first as brakeman, then as conductor, as long as the narrow-gage operated, and finally passed away on the very day it folded up! Whether or not Jim Shull consciously identified his life with that of the "Waynie" is a topic for speculation, but grief over its fate was more than he could bear. On June 10, 1929, when the last narrow-gage train was filling up with passengers at Waynesburg, Pa., the aged conductor showed up as usual to collect fares, but declined to go aboard. A brakeman took his place. Left behind, Jim watched the little engine and cars round a curve. Then he slumped to the platform and expired of a heart attack. That train, like the grandfather's clock "stopped, never to run again, when the old man died."

A common custom in bygone days was to lay on trainmen's graves floral tributes shaped like a broken wheel. This practice originated in 1887, when "Long John" Simpkins, conductor on a Colorado& Southern coal run, gave his life to keep a string of runaway cars from crashing into a passenger train, and a shattered brake wheel was found clutched in his stiff hands. The wheel was buried with him. Railroaders who attended the funeral commented that if this grim souvenir did not give John a clear board through the Pearly Gates, they themselves would stand a mighty slim chance at the end of life's run.

Here are a list of some more railroad superstitions:

- If a locomotive was involved in any fatal accident of its crew or passengers, it was considered a "hoodoo locomotive." Engines would be renumbered to "lift the curse."

- Railroaders believe accidents come in sets of three.

- The month of September is believed to be jinxed for railroaders.

- A white cat near the track is a bad sign; a black cat looking at the train as it passes is even worse.

- You will have a safe journey if you see two chickens fighting as you are leaving.

- Sweeping out a caboose after dark brings bad luck.

- If you count the cars of a passenger train, you will hear news of a death.

- To ward off unwelcome advances from a suitor, tie one of his socks to a freight train.

- If you want to get rid of your boyfriend, put one of his socks on the tracks. When a train runs over the sock, he will leave the same way the train went.

- It was believed that turning a locomotive into the sun would bring bad luck.

- On a roundhouse turntable, locomotives are only turned to the right; turning one to the left is bad luck.

- Leaving either the locomotive or caboose coupler open meant bad luck for the trip. Others believe that leaving the front engine coupler open will "catch" good fortune.

- Railroaders often carry good luck charms in their pockets to bring good luck on the trip.

- Beginning a trip on a Friday is bad luck; beginning on Friday the 13th is extremely bad luck.

- Taking a new engine out of the shop or starting a new route on a Friday is bad luck.

- Striking your foot with the switch frog while switching track is bad luck.

- Stepping on a rail will bring bad luck for the journey.

- Stumbling over a rail will bring misfortune; to avoid bad luck, go back and step again safely.

- Cross-eyed men (or women, in the case of passengers) bring bad luck.

- Counting the cars of a freight train as it passes will bring you good luck.

- A locomotive numbered 9 or 13 is bad luck. Number 13 locomotives on the Cairo & St. Louis, Iowa Central, Kansas

Pacific, Cairo & Vincennes, and St. Louis, Kansas City & Northern Railroads all were believed to be cursed (along with the No. 113 on the Illinois Central).

- Engines numbered 666 were a bad omen; the locomotive was doomed to bring death or grief to the crew. Engines with numbers adding up to nine are also said to be "hoodoo engines."

# Chapter Twelve: Conductors on the Screen

Throughout the history of movies and television, the railroad has served frequently as a backdrop for numerous productions. In some features, the train is the central theme with the conductor playing a prominent role. When thinking of the conductor character in the movies, we are usually drawn to the children's fantasy characters found in Shining Time Station and the Polar Express. Conductors, however, are featured in many action movies, dramas, and horror films in heroic and not so heroic roles. The following is a list of some memorable (and not so memorable) movies in which the conductor was prominently featured:

*Night Train 2009*

A train conductor (Danny Glover), a pre-med student (Leelee Sobieski), and a fast-talking salesman (Steve Zahn) find their greed clashing with their better judgment after stumbling into a fortune in diamonds aboard a speeding locomotive, and attempting to figure out a way to keep their discovery a secret. A nameless passenger has died en route to an unknown destination. His only possession: a mysterious box containing a valuable treasure. With that amount of money, the conductor, the student, and the salesman could all live comfortably for the rest of their days. But all three would rather have the fortune to themselves, and as the train races toward its destination, temptation, betrayal, deceit drives them each to irrational extremes in a paranoid bid to outsmart the others. Little do these three desperate souls realize that there's something sinister to this innocent looking box, and they might not find out what it is until it's already too late.

*Rails and Ties 2007*

Kevin Bacon stars in the story of a mentally ill, single mother who takes illegal drugs and is unable to care for her 10-year-old son Davey. Driven to despair, she decides to commit suicide by driving a car on to a railway track, taking her Davey with her. As a train approaches, Davey tries in vain to drag her out of the car, himself jumping clear just in time. Two train crewmen, Tom Stark and Otis Higgs, seeing the car on the tracks ahead, argue about whether an emergency stop will derail the train or not. However, the train hits and kills the boy's mother. Subsequently, the railroad company calls for an internal inquiry and suspends the two drivers.

*Come Together 2016*

Adrien Brody plays a character named Conductor Ralph, who has to inform passengers on Christmas that challenging weather conditions and mechanical difficulties have delayed their train nearly 12 hours.

*Midnight Meat Train 2008*

This classic proves that the conductor is not always a helpful, loveable character. In this classic, Leon (Bradley Cooper) is a photographer who somehow stumbles upon the "Butcher", a meat packer by day who murders unsuspecting victims on a train at night. Leon eventually battles the butcher on the train and emerges as the victor. With the butcher's death, the conductor of the train enters the car, advising Leon and his girlfriend Maya to "Please step away from the meat." With these words, the true purpose of the underground station is revealed, as horrible reptilian creatures enter the car, consuming the meat to which they have been delivered. The conductor explains to Leon that the creatures have always existed below the city, and that the butcher's job was to keep them satisfied by feeding them every night. The conductor then forces Leon to watch as he kills Maya with one of the butcher's knives. When he is done, he picks up Leon, and with the same supernatural strength as

121

the deceased butcher, rips out Leon's tongue, throwing him to the ground. He tells Leon that, having killed the butcher, he must take his place. In the final scene, the police chief hands the train schedule to the new butcher, who wears a ring with eight stars. The killer walks onto the midnight train, and turns his head to reveal that he is Leon, ready to go on his nightly slaughter

*Unstoppable 2010*

With an unmanned, half-mile-long freight train barreling toward a city, a veteran engineer (Denzel Washington) and a young conductor (Chris Pine) race against the clock to prevent a catastrophe.

*Emperor of the North 1973*

It is during the great depression in the US, and the land is full of people who are now homeless. Those people, commonly called "hobos", are truly hated by Shack (Ernest Borgnine), a sadistical railway conductor who swore that no hobo will ride his train for free. Well, no-one but "A" Number One (Lee Marvin), who is ready to put his life at stake to become a local legend - as the first person who survived the trip on Shack's notorious train.

*The Fighting Sullivans 1944*

This true story chronicles the lives of a close-knit group of brothers growing up in Iowa during the days of the Great Depression and of World War II and their eventual deaths in action in the Pacific theater. The story features Thomas Mitchell as their father, a freight conductor who has not missed a day of work in thirty years.


*Illusive Tracks 2003*

On a train in 1945 a train conductor is in charge of a motley bunch: a failed author who means well but creates chaos; a soldier who is actually on the wrong train; a doctor who wants to murder his wife; a gay man who hates men; and two nuns with religious doubts.

# Chapter 13: Conductors in Song

*Please Mr. Conductor, Don't Put Me Off:* Originally written in 1899, this song won a degree of popularity when it was recorded by the Everly Brothers

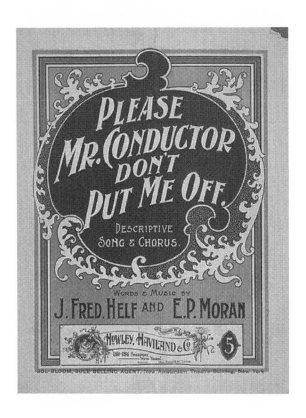

The lightning express from the depot so grand
Had started out on its way
All of the passengers that gathered on board
Seemed to be happy and gay
But one little boy who sat by himself
Was reading a letter he had
You could plainly tell by the look on his face
That the contents of it made him sad
The stern old conductor then started his round
Collecting tickets from everyone there

And finally reaching the side of the boy
He gruffly demanded his fare
"I have no ticket" the boy then replied
"But I'll pay you back someday"
"Then I'll put you off at the next stop we make"
But he stopped when he heard the boy say
Please Mr. Conductor
Don't put me off of this train
The best friend I have in this world sir
Is waiting for me in pain
Expecting to die any moment sir
And may…
Please Mr. Conductor
Don't put me off of this train
The best friend I have in this world sir
Is waiting for me in pain
Expecting to die any moment sir
And may not live through the day
I wanna reach home and kiss mother goodbye
Before God takes her away
A girl sitting near was heard to exclaim
If you put him off, it's a shame"
Taking his hand, a collection she made
The boy's way was paid on the train
I'm obliged to you miss for your kindness to me
You're welcome, " she said, never fear
Each time the conductor would pass through the car
The boy's words would ring in his ear
Please Mr. Conductor
Don't put me off of this train
The best friend I have in this world sir
Is waiting for me in pain
Expecting to die any moment sir
And may not live through the day
I wanna reach home and kiss mother goodbye
Before God takes her away

## Mr. Conductor Man 1932: Big Bill Broonzy

I got up this morning : hear the train whistle blow
Lord I thought about my baby : I sure did want to go

Lord I grabbed up my suitcase : I *dropped it on the floor*
I could see the conductor : he waving his hands to go

I said Mr conductor man : I want to talk to you
I want to ride your train : from here to Bugaloo

I'm leaving this morning : man I ain't got my fare
But I will shovel coal in your engine : till your train get me there

Crying please Mr conductor man : please take my last thin dime
Lord I got a woman in Bugaloo waiting : man I can't lose no time

When the bell started ringing : conductor hollered all aboard
Lord I picked up my suitcase : start walking down the road

I'm leaving this morning : I sure don't want to go
Lord and the woman I been loving : she don't want me no more

## Chapter Fourteen: Becoming a Conductor: Getting Started

If after reading this book you feel that it's your calling or your passion to become a train conductor, then congratulations. You will be embarking on a celebrated journey toward a career characterized by freedom, leadership, and respect. The following chapter will help guide you in starting down this path. If, on the other hand, you're still a bit uncertain. Ask yourself the following questions to better assess if the life of the conductor is the one for you:

- How motivated am I to work when I don't have someone watching over me?

  - This one is key. As a freight or passenger conductor, you are the authority on your train. Of course, that doesn't mean you're top dog. You still answer to a train company. However, when you're riding through the countryside, they won't be watching your every move. Some people find it challenging to work in these conditions. They are motivated by the presence of authority and the knowledge that if they aren't working, someone will know. A good train conductor is self-motivated and enjoys doing the job right with or without oversight.

- How well do I work with people? Do I feel I can manage them with little conflict?

  o This is important whether you are on a passenger train or a freight train. It goes without saying that passenger trains will have you surrounded by crewmen and working with people at every turn. However, freight trains require a degree of leadership as well. When you're not coasting along with your engineer, you're shunting trains with the help of a yard crew. It's important that you feel comfortable being the authority in those situations.

- Am I the sort of person who sticks to a schedule or do I prefer a more freeform existence?

  o If you aren't hopping out of bed at five on the dot every morning, don't fret. That doesn't make you a poor candidate for a conductor. This is more a matter of how organized and structured you are in your working life. Do you get to meetings on time? Do you prefer an unstructured work day where you can address whatever task moves you? Although there is a great deal of freedom in the life of a conductor, that freedom is on a schedule.

- Am I a patient person or am I prone to getting frustrated easily?
  - Conducting on either freight trains or passenger trains requires patience in more ways than one. First, there's the obvious patience that you may need to have with passengers who could, at times, be upset or demanding. However, there's also a degree of patience required when traveling long distances for long hours. A good conductor is one that stays focused and avoids restlessness.
- Do I prefer positions where I can multitask or would I rather focus on one thing until it is complete?
  - Of course, it is important to see things through when you are a conductor, as in all things. Trains need to get to their destinations on time and sometimes the conductor needs to prioritize certain tasks above all

others, making sure they are completed before anything else can happen. This is especially true when it comes to shunting, checking machinery, and similar tasks related to the safety and operation of the train. However, conductors also need to be effective multitaskers. They need to be able to handle passengers, paperwork, and crew while also keeping an eye out for safety hazards and communicating with stations. Even freight train conductors, who don't have to worry about the passenger aspect, must remain vigilant, communicative, and complete multiple tasks.

- Am I prepared to take on a great degree of responsibility?
    - As a conductor, what you do is very important. Getting to your destinations on time with the correct cargo (or passengers) can affect numerous other things. You can set the tone for people's entire vacations. You can also prevent the loss of millions of dollars by getting cargo where it needs to go on time. And the cargo isn't the only valuable thing a

conductor is responsible for. The train itself is a very large and very expensive piece of machinery. As the conductor, you are top dog on the train and therefore responsible for it. Accidents involving trains can be very serious and it is up to you to adhere to the necessary safety precautions.

*Deconstructing the Job Posting*

In truth, we're only going to concern ourselves with one key portion of any posting: qualifications. This, after all, is the section that will help you understand what you need to know before you even attempt to land a job as a conductor. Note that the path to becoming a fully realized train conductor varies depending on whether you're looking to be a freight or passenger conductor and the specific company you are applying to work for. Many companies have their own "paths" for conductors. We'll cover some of these later in the chapter. Here, we simply want to gain an understanding of the general skills and education that most companies want to see in a prospective conductor.

*Minimum Qualifications*

Before you can even consider applying for a job as a conductor, you need to meet certain basic requirements. First, you must have a high school diploma or equivalency (like your GED). You must be at least eighteen years old. Finally, you must have a driver's license. All American train companies will look for these three things and there are no exceptions.

Some train companies will lay out additional requirements. If you are applying for a job as a passenger train conductor, you may be expected to have a safe work record and satisfactory or stellar attendance at your previous job. They'll likely ask your previous employers for information regarding these attributes. You may also be asked to have some specific experience exhibiting leadership or experience with customer service. Other companies may ask that you have a certain level of education (either on the job or in higher education). They might ask that you have at least two years of verifiable work experience or two years at an accredited college or university. Finally, many companies will require that you have no driving violations involving substances on your license within the past three years or at all.

Read any job descriptions you find carefully. Anything listed as a minimum qualification is a deal breaker. In other words, don't waste your time applying if you lack any of them. You could be a stellar candidate in every other regard, but they will still dismiss your application. That being said, don't lose hope. Different train companies have different standards and if you work at it, you'll find the one that's right for you.

*Preferred Qualifications*

These qualifications are not deal breakers. It is always good to be able to match some of these standards. However, if you don't and you think you might otherwise be a strong candidate, apply anyway. These standards are mutable for a reason. Train companies understand that good conductors can come from a variety of places and they don't want to limit their options too much.

Standards like these might include having experience working outside, on-call, with shift based schedules, or with heavy equipment. Some companies may also ask that you have some mechanical experience or have spent some time in a supervisory role. In short, it's helpful for these companies that you have spent some time on relevant tasks or displaying desired qualities.

However, they hold to the fact that they are willing to train if you don't have this coveted experience.

*Physical Requirements*

Most train companies will have physical requirements. This is especially true if you are applying for a job as a freight train conductor. In most cases, they expect you to be able to lift a minimum of fifty pounds and up to eighty pounds on occasion. However, certain companies might require more if they know the role calls for it. Never apply to a job if you can't meet the minimum strength requirements. You are only setting yourself up to fail and, possibly, to get hurt.

Strength isn't the only physical requirement that train companies hold to. Nearly all companies will have a color vision requirement. In short, if you are colorblind, you likely won't make it as a train conductor. Conductors need to be able to read signals clearly and this often involves color coding. While it may be frustrating that your colorblindness could be keeping you from your dream job, never attempt to mask it. Doing so will only put yourself and others at risk. Along the same lines, some companies may have straight vision tests. If you wear corrective lenses, don't fret. Unlike with pilots, you are not expected to have uncorrected 20/20 vision.

Some companies ask that you be able to walk long distances over uneven terrain. The ballast on train tracks may look tame from a distance, but it can be treacherous, especially in poor weather. It is important that a conductor be sure footed lest he or she suffer a fall and hurt an ankle or worse. Notice that they walk for long distances as well. It is one thing to be able to walk on irregular terrain for a short time and entirely another to do so for an hour or more. The skill is as much about strength as it is about balance.

When covering the various responsibilities of the conductor, we talked about shunting, coupling, and uncoupling train cars. These tasks often require conductors to grab onto a ladder on the outside of a train car and hold on while the car is in motion. Obviously, if you're new to the profession, you can't know with certainty if you are physically up to the task of riding on the outside of train equipment for extended periods of time. However, you want to think critically about your abilities and answer honestly about whether or not you think you can handle this type of work.

Another physical challenge that conductors face is working in cramped and confined spaces. This is especially true for freight train conductors who spend long periods of time in small train cars and relative isolation. This might seem like a no brainer and you're likely to tell an interviewer that you're up for it, but you might be surprised how physically and mentally taxing this can be. Imagine working in an elevator all day or being confined to a cubicle for hours on end. If you can't even handle an hour commute to and from work in your car, you might not be cut out for this.

Finally, employers will ask that you can work in all kinds of conditions, meaning rain, snow, fog, freezing temperatures, and temperatures cresting one hundred degrees. Conducting is not, strictly speaking, an outdoors job. After all, you spend much of your time confined to the train. However, there are significant portions that are outdoors. You may have to contend with the elements and extreme temperatures when shunting train cars, ushering passengers on and off train cars, or loading and unloading cargo. Remember that your job involves keeping to a strict schedule. Therefore, there's no opportunity to flinch at poor conditions.

*Drug Testing*

It's not enough to apply and wait for a yes. Companies have other requirements that could stand between you and your goal. During the pre-employment process, you'll likely be asked to submit to a drug test. In fact, you're probably going to be drug tested on a regular basis. You're going to be in charge of a very valuable and dangerous piece of machinery. There can be no room for error and companies are not inclined to take any risks.

*Getting Your Foot in the Door*

If you meet the minimum and preferred qualifications put forth in a conductor job description, you are by no means guaranteed an interview. Note that the bare minimum requirements for becoming a passenger or freight conductor are relatively low. In some ways, it almost looks like an entry level position. However, it is quite the opposite.

Although having a high school diploma or equivalent, having a driver's license, and passing a drug test are likely to qualify you for a training program, you still have your work cut out for you before you actually find yourself conducting any trains. Train companies value experience, be it with machines, management, heavy lifting, customer service, or all of the above. If you lack any work experience that can translate to your sought after position, you'll likely get passed up for a training program.

If you are not accepted into a training program, don't lose hope. Train companies value their hard-working employees. As such, they encourage upward momentum within their ranks. It is very common for conductors to get their start as brakemen or switchmen. As you gain relevant experience, you can move up the ranks. This is the longer route to your desired destination and that can be understandably frustrating. But there is something to be said for gaining experience at different jobs within the railroad industry. Doing so will help you appreciate train operations at a variety of levels.

If you manage to get accepted into a training program, you might feel like you're home free. After all, many of these programs are paid and it could feel like you're an official conductor already. However, assignments are slight and you'll still need to distinguish yourself in the training program if you want a guarantee that you'll be conducting trains within the next several months. Some railroad companies, like Union Pacific, will decide during training if you are cut out to be a conductor right out of the gate or if you are better

suited to a role as a brakeman or switchman for the time being.

# Chapter Fifteen: Becoming a Conductor: Getting Hired

Of course, the application is only a portion of the battle. Standards for conductors are high and you'll be going up against numerous applicants who are just as qualified as you. In other words, there is a big difference between getting started and getting hired.

*Railroad Pre-Employment Exams*

Let's say you've applied for a conductor position and the company in question has shown interest. What happens next? In most other jobs, you might be expecting an interview, but don't count your chickens just yet.

You will first be invited to an employment fair or similar function where you will need to qualify for an interview by passing a test (or tests) that measure your aptitude for certain skills. Of course, there is no hard and fast rule that train companies in the United States must administer pre-employment tests. However, most do and with good reason. They are often saturated with applicants and pre-employment tests are an effective way to narrow down the field.

Don't be alarmed. These tests are not aiming to assess knowledge and skill when it comes to specific conductor duties. Companies like BNSF Railway understand that they have job training for that sort of thing. Instead, they look for more general attributes. The BNSF test, like others, is comprised of mathematics and English. More specifically, it seeks to test you on things such as mechanical reasoning and communication. Many of these tests may also employ personality items. In other words, they look for applicants whose temperaments and problem solving skills are conducive to this line of work. These companies look for applicants with a baseline understanding not of conducting itself but of the essential skills upon which a good conductor is built.

*Types of Exams*

The test you take at a hiring fair will depend not only on the company to which you've applied but also whether you are looking to be a passenger or freight train conductor. The following tests are a sampling of what you might encounter. If you're unsure of what tests you will be expected to take, take some time to prepare for all possibilities.

- Math
    - The math you are expected to perform generally falls under the purview of arithmetic and algebra. You'll likely be asked to complete two step equations

141

involving multiplying, dividing, adding, and subtracting decimals and fractions. If you are applying to be a passenger train conductor, you can also expect to be dealing with percentages and converting decimals and fractions into percentages. Often, these tests are timed, so being able to do mental math puts you at a huge advantage.

- Mechanical Reasoning
  - Mechanical reasoning doesn't expect you to be able to complete any complex specific tasks when it comes to machines. Rather, it focuses on cause and effect and motion when it comes to mechanics. Expect questions to address levers, pivots, and gears and how they work with one another.

- Verbal and Deductive Reasoning
  - This test focuses on logic. Questions will likely ask you to draw logical conclusions based on sentence structure, make appropriate analogies, or deduce facts based on given criteria.

- Reading Comprehension
  - You've likely already experienced this in grade school. Questions reference a passage and ask you to summarize, deduce, or infer based on the information in the passage. You may be expected to identify main ideas or use context clues to infer the meaning of obscure words.
- Strength Test
  - This test takes you away from the pencil and paper and has you perform actual feats of strength. Of course, this is no strong man competition and you are not expected to exert yourself to the point where you might incur injury. The strength test simply seeks to assess how well you are able to perform relevant job duties that may require lifting and moving.
- Concentration Test
  - Focus test might be a more apt name for this activity. The concentration test seeks to test your ability to stay focused on a task and perform it accurately at

high speeds. You'll likely be subject to the Group

Bourdon test. This test, a derivative of the Bourdon-

Wiersma test, is a psychometric test that is very

popular for testing train drivers in the UK. However,

some US companies also employ the test (or similar)

when assessing the skills of prospective conductors.

The test presents individuals with collections or dots

or various shapes. The individual will then be asked

to perform a task, like identify every collection

containing three dots or cross out every iteration of a

particular shape. They are timed while they do it. The

test looks for quickness and accuracy.

- Personality / Behavioral
  - This test usually takes the form of role-playing. If you
    
    are subject to this test, you will be presented with
    
    various hypothetical situations and asked to respond
    
    in the most appropriate manner. This test typically
    
    involves manager / employee or conductor /
    
    passenger relationships. As such, it is likely only to

come up if you are applying for a job as a passenger train conductor.

- Safety and Common Sense
    - o It's all in the name for this test. You will cover a lot of on the job safety standards during your training. For this test, you're mainly expected to display sound judgement in common sense situations. This should be the easiest test to take without preparation.

- Visual Reasoning.
    - o Conductors need to keep their eyes wide open. Your sight is hands down the most important sense you'll use on the job. Some train companies may ask you to take timed tests involving videos with accompanying questions. This test is less common but equally as important if it comes up.

*Passing with Flying Colors*

First of all, "passing" is a relative term here. This isn't high school and you're not aiming to score a 60% or above. Your success on these tests is determined partly by the baseline standards of the company in question and partly by the performance of your peers. You can't know what kind of competition you're up against, so you can't know the bare minimum you're aiming for. All you can do prepare and do the best you can.

There is a wide market for classes, books, and other services to help prospective applicants prepare for railroad pre-employment exams. And who knows, maybe investing in one of these tools will help you unlock that coveted interview. However, if you're not interested in these products or able to shell out the money for such an advantage, consider the following tips to help guide you toward success:

- Brush up on your basic math.
  - You don't need a fancy book to do this. Online math games and quizzes can help you practice. You can also find affordable math guides for standardized tests at your local bookstore. Focus on
- Read.
  - You're doing it right now. The best way to brush up on reading comprehension is to read a lot. If you want to practice comprehension questions, consider

downloading an SAT prep app. There are many out there for free.

- Do strength training.
  - ○ Some railroad companies have applicants take strength tests. Partly, they want to ensure that these applicants meet basic heavy lifting requirements. However, it never hurts to set yourself apart by going the extra mile.

- Practice concentration tests.
  - ○ Take the Group Bourdon or similar psychometric tests online and challenge yourself to do better each time.

- Role-play.
  - ○ You may not know exactly what hypothetical situations the test in question will throw your way, but you can practice likely scenarios. Have your friends or family toss out potential questions or conflicts that you may encounter on the job.

- Do the leg work.

o   Most companies advertising conductor positions promise that you don't need to be a railroad expert, or even have any experience working on trains, to become a conductor. That being said, it doesn't hurt to do a little background reading. Learning a bit about the job responsibilities, safety standards, and common challenges, will help you better prepare by giving you a slightly better sense of what answers they might be looking for.

If you don't do well the first time, don't fret. There are many train companies out there and you can always try again. Don't forget to talk to the representatives at the testing center or job fair. They may be able to tell you where you excelled and where you need to practice. Also, understand that the test is only one piece of the puzzle. You could do reasonably well on all of the exams and still not get into a training program if you don't meet other requirements.

*The Interview*

Based on your performance in the pre-employment exams, you could be invited to a one-on-one interview. This will likely happen during the same job fair in which you took the test. Therefore, you'll want to dress your best and prepare for the interview in advance in case you get one.

One of the best ways to prepare for a job interview is by familiarizing yourself with common interview questions and practicing your responses. Train conductor interview questions can be boiled down to four main categories. Each of these categories serves a different purpose in the interview.

- Situational Questions
  - These questions focus on experiences you've had in the past and how you handled yourself. They are intended to give the interviewer insight into how you've historically managed conflict, stress, and other workplace challenges. These questions may also seek to understand times when things went poorly to better assess how you learn from past mistakes. Examples include:
    - Tell us about a time you had to respond to an emergency.

- Tell us about a team experience you found disappointing and why.

- Tell us about a time when you had to handle interpersonal conflict.

- Tell us about a time when someone was doing something unsafe. How did you respond?

- Understanding of the Role

  o These questions require you to display some knowledge of train conducting. In short, the interviewer wants to make sure that you know what you're signing up for. Additionally, they want to make sure that you have an accurate understanding of the qualities and skills they are looking for. Examples include:

    - What skills do you feel are necessary for train conductors?

    - What educational preparation do you feel is necessary for train conductors?

    - If you were hiring a person for this job, what would you look for?

- What are the three most important attributes for a train conductor to have?

- Fit Questions
  - In other words, are you the right fit for the role and the company? No matter the job in question, applicants will all too often just try to give the answers thy think the interviewer wants to hear. They have a one-track mind and getting a yes from recruiting is all they care about. In reality, this portion of the interview benefits you as much as it does the organization. After all, you have certain needs, desires, and general preferences informing your decision to make a career change. Be honest with yourself and your interviewer about those questions to make sure that this is truly the correct fit. Playing pretend will only give the interviewer the wrong idea and set you up for a job that doesn't really make you happy. Examples include:
    - When are you most satisfied in your job?

- Do you prefer to work independently or on a team?

- How do you keep yourself organized?

- What do you expect from this job? Salary? Benefits?

- What kinds of situations do you find most stressful and why?

- Hypothetical Questions

  - During your interview, you may face some hypothetical scenarios that you need to respond to. This is similar to the Personality / Behavioral or Safety and Common Sense tests detailed above. If you are required to take these tests, you will likely not receive any hypotheticals during the interview. Otherwise, expect a few as they help give the interviewer insight into your reasoning and problem solving skills. Examples include:

    - If you saw someone being unsafe at work or breaking the rules, how would you handle it?

- If a superior asked you do to something that was unsafe or against the rules, how would you handle it?

- If there was a conflict between two passengers how would you handle it?

You might find yourself up against questions that ask you to reflect on poor experiences or past mistakes. Such questions might ask you to reflect on a time when you tried to resolve a workplace conflict and it went poorly. They might ask you to list three positive qualities that you feel you lack. And of course, there's that most dreaded interview question asking you to identify your greatest weakness. Too often, applicants think there's some trick to answering these questions. They try to come up with a response that is really a veiled strength or success. The truth is that interviewers can and will see right through that.

Try to think of these questions not as opportunities for you to humble brag your way through the interview. Rather, take a moment to actually reflect on the question and answer it honestly. This is especially true for those questions that ask you to reflect on failed experiences. Remember that the point is not to make a failure look like a success but to show that you can accurately identify a failure and learn from it. Employers don't expect perfection; they look for honesty, integrity, and a willingness to grow.

Of course, this isn't the only interview faux pas you might commit. There are a variety of things that could be looked on unfavorably by your interviewer. Pay attention to the following "don'ts" so you know what not to do.

- Don't be late.
    - This is bad policy for any interview no matter what you're applying for. However, it is especially bad when it comes to train conducting. After all, punctuality is one of the staples of the position.
- Don't dress down.
    - This is especially true for those applying to passenger train conductor jobs, as they will have more stringent uniform requirements. However, you should dress in a suit no matter what type of conducting job you're applying for. You're likely to see other applicants in polos or even more casual dress. Their argument is

that it is a physical job and doesn't require business attire. Ignore them. An interview is always an occasion for business dress and it is always better to overdress than underdress. It tells the interviewers that you are responsible and serious about the role.

- Don't be restless or impatient.
  - o Interviews can be as short as half an hour or as long as two hours (or longer). The interview will take as long as it takes. If you behave in a restless manner, you will effectively be telling the interviewer that you don't value that position and are unwilling to give it the attention it deserves. What's more, flexibility and a willing to adhere to odd hours is an integral part of the role. Acting like you have other places to be suggests that you don't have either or those qualities.
- Don't be timid.
  - o This is a tough one if you are naturally inclined to be introverted. However, the interview is your opportunity to sell yourself to the company in question. Even more important is the fact that a timid

conductor will not be a successful one, whether applying to a freight or passenger train position. In either case, you are the authority on a train, no matter how many people you interact with daily. That means you need to carry yourself thusly. Act like you belong in that role and can be trusted with that responsibility.

- Don't get defensive.
  - On the other side of being timid, there's getting too aggressive. Interviews are stressful and it can be easy to get defensive if you feel you're not coming across well or if the interviewer asks you a probing question that you're not prepared for. This is always a bad thing in interviews. However, conductors are expected to be skilled at handling high stress situations and diffusing conflicts. If you can't handle the interview, you can't handle being a conductor.

*Conditional Employment*

After completing a hiring session, you will play the waiting game for a while. Eventually, if the company thinks you are a good fit, you'll receive an email or phone call congratulating you on your employment and informing you about next steps. Sometimes this employment is called "conditional employment." In other words, you are hired under the expectation that you will meet certain requirements.

If you have been offered conditional employment, expect to go through a series of pre-employment steps even now as you wait to begin conductor training. For example, you might need to get a physical, pass a background check, and have your former employers verify your employment with them. The company assumes that you will meet these standards and therefore, tells you that, provided all of the Ts are crossed and Is are dotted, you are golden.

*Setting Yourself Apart*

We've said time and again that the only set in stone education requirement for conductors is a high school diploma or equivalent. That being said, there is always something you can do to make yourself a more appealing candidate, and we're not just talking about the above tips on preparing for the hiring fair.

You may not need a higher education degree, but that doesn't mean you can't get one to really set yourself apart. Many community colleges offer degrees in subjects like Railroad Science. If you find yourself struggling to land the job you want, consider spending a couple of years going through one of these programs. It will make you look knowledgeable and serious.

You could even go all out and get a degree in something like mechanical engineering. Although if you go so far as to get a four-year degree in such a coveted area of expertise, you may find other high paying jobs in your future.

Finally, if you have the opportunity to work on trains prior to applying for a conductor position, you may find this background knowledge can help give you a boost. However, if your passion is conducting and conducting only, don't sweat it. Plenty of applicants will be coming in just as green as you.

## Chapter Sixteen: Conductor Training

If you browse corporate websites or job postings, you may notice the promise that you don't need experience with trains to qualify for a position as a conductor because you will be thoroughly trained. As was previously stated, this is only partly true. Some train experience is valued and you may find yourself starting at the bottom and moving up. However, the later part is all true. Conductor training is extensive and comprehensive. While there you will learn about everything from basic policies and procedures to handling machinery and filling our reports.

Whether you're just being hired into the role or you've worked your way from the bottom up, you'll likely be put through some degree of conductor training. The extent of this training varies dramatically from company to company. Union Pacific promises fourteen weeks of training (roughly three months). Canadian Pacific, on the other hand, boasts training that is between four and six months. CSX Corporation promises a very wide range. They require four weeks of training at a central facility with another eight to twenty-two weeks of on the job training (roughly 2-6 months in total). Whatever your prospective employer, you can expect to be subject to a least a couple of months of straight training. In nearly all cases, this training is paid.

*Sample Schedule*

Although the training is somewhat varied, you can count on certain things being covered. The following is a sample schedule for a conductor training and may give you some idea of what might be in store.

- You'll spend a day or three filling out forms. Boring though it may be, it will help get you set up for compensation, pension, insurance, and more.

- Training generally starts with an HR style intro to the company. You'll likely be met with one to three days of platitudes about what the company represents and how they work to serve the public. Don't worry, the nitty gritty stuff (and the stuff you'll use most) is coming.

- Sometime during those first weeks, you'll meet union reps and get established in your local union chapter.

- During that time, you might be measured for a uniform. However, if your training is partly used to determine aptitude and may place you as a brakeman or switchman to start, you're not likely to get a conductor's uniform until later.

- After covering the company basics, you'll be met with some more intense on the job training. Here you will learn the ins and outs of train conducting and related skills. You'll likely work with some heavy machinery and will be briefed ad nauseam about safety procedures.

- When you're training is over, you'll be given your assignment. In some cases, that may mean learning whether you are going to be a full-blown conductor right off the bat or going to start paying your dues in another role. If you entered a training program knowing you were already guaranteed a position as a conductor, you'll learn where you're assigned and what your shifts are.

- On the job training may continue for another several months. During much of this time, you'll not be managing trains alone but rather working with other conductors, as you learn by doing.

*Homework*

Maybe you thought you left homework behind when you finished high school. Think again. Conductor training isn't like on the job training for your average office position. It is lengthy and intense, and it comes with high expectations. Your conductor training program will almost certainly have work for you to take home at the end of most days.

More often than not, this work will involve a lot of reading. However, you may sometimes be expected to take home practice quizzes or tests. If your homework does involve a lot of reading, don't shirk it. You will certainly be quizzed and tested during your training sessions. And make no mistake; the railroad representatives training you will take them seriously. If you consistently underperform on these test, you might be asked to leave the program. Don't let that scare you though. If you're serious about being a conductor, you'll want to learn the trade. That means working hard.

*Other Tests*

Not all of the tests you take during training will be pen and paper. Many training tests will involve physical activities. After all, it's one thing to know the ins and outs of the job and another to be capable of performing the necessary duties. One such test is called the hang test. This is a popular test among conductor training programs but your training program may or may not include it. This test requires trainees to hold onto a steel ladder for a specified period of time (often one to two minutes) while performing another function with a free hand, usually making signals. In most cases, you will pass even if you get the signals wrong. The trainers are primarily looking to determine how physically capable you are of multitasking while holding on to a train side ladder. If you consistently fail to hold on, you will likely be asked to leave the program.

At the end of the training, you will be subject to a large cumulative exam. This test will cover everything you've learned about conducting and determine how prepared you are to move forward.

# Chapter Seventeen: Where to Apply

Being familiar with the application process and the various demands that may be placed on you is only half of the battle. To say that there are dozens of railroad companies in the United States is to put it very mildly. In fact, there are dozens of railroad companies in the United States that start with "A" to say nothing of the rest of the alphabet. And that's only part of the story. Canadian rail companies have lines that run through various parts of the US and many American rail workers have gone on to work for them. Basically, you've got your work cut out for you if you think you're going to apply to every job under the sun. It's time to think hard about the kind of conductor you want to be and the kind of company you want to work for.

*Ask the Obvious Questions*

We'll begin by offering some guidelines for finding the right fit. The following questions should hopefully already have occurred to you, but it's worth emphasizing the importance of knowing what you want and going after it. If you're looking to become a conductor, then you are looking for a career change. Don't settle now.

- Passenger or Freight?

- Let's start with an easy one. These jobs have some very dramatic differences. For more information on what distinguishes a passenger conductor from a freight conductor, visit the chapters on Types of Conductors and Responsibilities.

- Where do you want to work?
  - Are you only looking for work based out of your home town or city? Are you willing to travel to a new city for the right job? If so, which cities would you prefer and which are deal breakers?

- What is the lowest salary you are willing to work for?
  - While there is some consistency across the industry, different train companies may have varying salary schedules. This is especially true for smaller companies that may not be able to compete with larger ones.

Most freight companies are very similar from an employee perspective. They follow the same industry regulations and have similar contracts with local unions. Passenger companies might vary a bit more in terms of length of shifts and the number of crew members under you.

*Highlighting the Biggest America Has to Offer*

Obviously, we can't do justice to every potential employer on the market. That would take an e-Book about ten times the size of this one. However,we can take a peek at the biggest companies in the United States and, therefore, the ones you are most likely to end up working for.

*Amtrak*

Amtrak is the public name for the National Railroad Passenger Corporation, and it's the most highly recognized passenger train company in America. Amtrak reaches not only forty-six states in the US, but also three Canadian provinces. With more than 500 destinations, it is an extensive railroad network that works hard to get people where they want to go.

Amtrak offers extensive benefits to its employees, including vacation, holidays, rail-privileges, and family and medical leave. Their health benefits cover straight health insurance, dental insurance, vision insurance, and more. Because they are such a large company, they are able to take care of employees in this way.

Despite Amtrak's many benefits, it does suffer from some issues in the satisfied customer department. Amtrak has an F rating with the Better Business Bureau largely due to complaints about delays and, overcrowding, and customer service. It also has a low rating with consumer affairs for much the same reason. That being said, there are plenty of people who ride the Amtrak rails on a regular basis and have no qualms with the company.

*BNSF Railway*

The second largest freight company in the United States, BNSF is a Class 1 railroad. This means that their annual revenue is consistently at least $250 million. This company is headquartered in Fort Worth, Texas, but BNSF lines run across twenty-eight states on the middle and western portions of the country and carry more than eight thousand trains. In other words, you don't need to be Texan to find work with them.

Another large operation, BNSF also takes care of its employees, offering competitive pay and benefits. They also promise significant opportunities for advancement. Although these are likely in line with the typical modes of advancement that railroad workers enjoy.

BNSF has an A+ rating with the Better Business Bureau due to a lack of response (no complaints and no reviews). However, before you start assuming that it is better than Amtrak for that reason, note that customers are far more likely to log complaints than employees and Amtrak suffers from an overabundance of customers. Still, it's worth noting that the official BBB safety record for BNSF is in tact.

*CSX Transportation*

Like the BNSF Railway, CSX Transportation is a Class 1 freight railroad company. It serves primarily the eastern third of the United States and reaching up into Ontario, Canada. CSX Transportation is a subsidiary of CSX Corporation, an American company whose primary business is owning shares in the real estate and railroad industries.

There are benefits and downsides to working for a subsidized company. Sometimes, knowing that there's a larger company out there interested in the success of your company can help foster a sense of security. Of course, this is less necessary with a company that's already as big as CSX transportation. Sometimes people dislike working for subsidiaries. They worry that the parent company has too many interests unrelated to their own.

In either case, CSX Transportation offers competitive benefits and pay and enjoys a mostly A+ rating with the Better Business Bureau.

*Kansas City Southern Railway*

This railroad company has the unique distinction among the top US railroad companies to service Mexico. Based out of Kansas City, Missouri, it is the smallest of the Class I railroads and serves the smallest number of states of the companies listed here. Despite its size, Kansas City Southern offers extremely competitive benefits, including health insurance, vision insurance, dental insurance, opportunities to buy stock at a value, and a 401k retirement plan. They have no visible rating with the Better Business Bureau, which could be considered as good as an A+ what with the lack of complaints.

*Norfolk Southern Railway*

Norfolk Southern operates exclusively on the east coast, competing directly with CSX Transportation and the Canadian Pacific Railway. Another Class I railroad, it operates out of Norfolk,

Virginia and runs through twenty-two states. The company claims to be the largest intermodal freight train on the eastern half of the continent. This means they carry cargo in large containers that can easily be moved from train to ship to airplane without having to handle the contents. Norfolk Southern offers competitive rates and salary. It has a mostly A+ rating with the Better Business Bureau and mixed reviews from employees on Glassdoor.

*Union Pacific Railroad*

This company has the distinction of being both the largest and one of the oldest train companies in America. It hauls freight from the middle of the country clear to the west coast. The Union Pacific Railroad started with the Union Pacific route, the western half of the Transcontinental Railroad. From there it grew into a massive corporation that hauls freight across twenty-three states using eighty-five hundred trains.

Union Pacific takes pride in its history and the role its original route played in shaping the country. Visit their website and you'll see that they treat their employees like a community of people working together to build a better future. Whether you buy into this or not, it's clear that they have a passion for motivating workers and a strong image to uphold. Therefore, they're likely to have tough standards and take care of their workers. As with the other freight companies on this list, they have an A+ rating with the Better Business Bureau.

*Why So Few Passenger Companies?*

It's true. Amtrak seems to have the unique distinction of being both a passenger train company and a major train company in the United States. However, that doesn't mean it's the only one out there. Plenty of smaller passenger trains still operate within cities and states. Several work somewhat longer routes, crossing state lines. Most of these are public organizations, funded and run by local government. For example, the Massachusetts Bay Transit Authority runs a commuter rail that travels between various points in northern, eastern, and western Massachusetts and Rhode Island.

Passenger trains will never the be money makers that freight trains are. Amtrak is unique among them, with 2.185 billion dollars in annual revenue in 2015. To put that in perspective, consider that freight companies like Norfolk Southern make tens of billions of dollars each year. In short, passenger train companies are rarely going to make as much as freight companies.

What's more, smaller passenger train companies will never be able to compete with Amtrak. Amtrak is a government mandated monopoly. It is a public company created by Congress in 1970 to be the sole passenger train operator on cross-country tracks. Prior to this, private companies shared the railroad and operated at enormous losses. Congress's act was intended to save the passenger train industry, recognizing that demand was not high enough for free market competition to work.

Of course, not everyone is happy about this change. In 2015, Kevin McCarthy, the House Majority Leader, released a statement to the House of Representatives appealing to them to end Amtrak's monopoly and the governments interference in the free market. He claimed that the government was not helping to save the railroads so much as preventing them from being able to save themselves.

Until such changes occur, Amtrak will be your best bet for passenger train jobs. And if you dream of traveling long distances as a passenger train conductor, it will be your only bet. Otherwise, look to your city's or state's local transit authority. You might also look to see if there are any private trains in your area. Many of these operate as attractions, like the scenic trains that wind through portions of the Rocky Mountains. They may also come with better, more reliable hours.

*Now What?*

Have a first choice? Make sure you have a second, third, and so on. In fact, don't hesitate to apply to just about any job that fits the bill. Do this by going to the railroad company's website. Somewhere on the landing page you should see a link that says something like "jobs" or "work here." These companies do most of their hiring from their websites and will make it very easy to navigate and apply to openings.

Your job doesn't stop with applying either. Check back frequently for more job postings and announcements. If you are serious about becoming a conductor, you'll check the postings every day. Some companies may even reach out to applicants through the system, so checking will not only keep you on top of new openings, but also prevent you from missing an important communication about a job you've applied to.

# Chapter Eighteen: Conductors, Engineers, and Your Future

Conductors and engineers are the two most prestigious and essential positions on any given train. Conductors manage the train's overall operations, while engineers actually operate the train. These positions are so essential that many freight trains function with only a conductor and an engineer on board. They are the only positions required to be on any operating train.

Although conductors are in charge of the trains and, therefore, hold a higher authority than engineers, the later is considered the "end game" job in the train industry. Just as brakemen and signalmen might eventually go on to become conductors, conductors often go on to become engineers. If you are interested in becoming a train conductor, then this role is likely in your future. So let's take a moment to learn more about engineers and why so many conductors end up driving trains.

*The Engineer in Depth*

An engineer's responsibilities are basically the same whether he or she is driving a passenger or freight train. The only notable differences involve likely hours and the obvious fact that passengers prefer lighter handling (if a freight train engineer hits the brakes a little too enthusiastically, the cargo isn't likely to complain). For the most part, however, driving a train is driving a train and that is all engineers really concern themselves with.

Of course, driving a train is not nearly as simple as it might sound. If one were to romanticize the role of the conductor, one might imagine a laid back individual pressing start on a train engine and leaning back to enjoy the countryside as it trundles by. The reality, however, is quite different. Engineers must be vigilant at all times and none of what they do is as simple as it first sounds.

For example, engineers are responsible for managing the speed of the train. But this doesn't mean braking and accelerating the way you might in a car. Trains travel at such great speeds that engineers must plan breaking far in advance. This means being intimately familiar with their route; they must be aware of signals, safety procedures, and even just the condition of the track. Additionally, they must work closely with the conductor and dispatchers to make sure that they are prepared for unscheduled stops.

In addition to monitoring speed and ensuring that the train brakes on time, engineers must stay on top of air pressure, battery use, and anything else that might interfere with the train's safe movement. They communicate primarily with the conductor and relay important information regarding train operation.

Sometimes the engineer's responsibilities overlap with those of the conductor. For instance, engineers may find themselves shunting and coupling train cars in the yard. They may also communicate directly with dispatch or address mechanical issues on the train. This is one reason why having had training and experience as a conductor first is so important.

Like a conductor, train engineers often have very irregular hours and long shifts (especially if they work on freight trains). Conductors who reached a certain level of seniority and enjoyed better, more reliable shifts as a result, shouldn't expect that when they transition to engineer. Once they become an engineer, they are a "rookie" again and will have to work harder shifts in the beginning. That being said, engineers are never expected to work more than twelve hours in a single shift. In fact, doing so would violate important national safety standards.

Driving a train is a hard life. Engineers are on call day and night. They are sometimes given as little as two hours' notice before they have to report for a shift. Nothing is safe, not weekends, not holidays, and not children's birthdays. However, if you've worked successfully as a conductor, you already know a little about this kind of spontaneous living and may be well prepared for the role.

*How Conductors Become Engineers*

Not only is it possible to transition from being a conductor to being an engineer, it is largely expected. For many reasons that we've already stated, conductors are well suited to being engineers. They are already familiar with the hard life of shift work combined with long hours. They understand the safety procedures and mechanical workings of the train. If they are freight conductors, they are likely also familiar with the relative isolation that train riding can bring.

Train companies understand this and, for that reason, they've made it very simple to transition into an engineering position (if not easy). Conductors who want to be engineers can submit paperwork requesting admittance to an engineering training program. This is also known as "bidding." Unfortunately for those who are new to conducting and want to make the switch, these bids tend to go by seniority. Train companies will consider skill and service, but ultimately everyone needs to wait their turn.

If you bid on an engineering program, make sure you're serious about the job. Your conductor position will not be waiting for you if you fail the program or change your mind and leave. These programs tend to last about five to six months and train companies can't afford to hold conductor jobs open for that long.

Conductors who are lucky enough to get into one of these programs can look forward to training as or more demanding as their conductor training. Engineer training is chock full of tests on mechanics, rules, and safety procedures. When they aren't studying and testing, prospective engineers will be practicing in simulators, upon which they will eventually also be tested. Standards for passing these tests are high. Sixty or seventy percent might have been good enough in high school, but here you'll likely need at least ninety percent to pass. It's a dangerous job and the train companies don't want to take any risks. Once this training is complete, engineers will have a federal license to operate a train.

Next, you'll begin your on the job training. This typically lasts about sixteen weeks. During this time, you'll be trained on a specific route and train. Different types of trains can handle very differently. It's important that engineers are comfortable with the type of train they will ultimately be responsible for. This time is also dedicated to helping them get comfortable with their route. Engineers need to anticipate stops and other potential hazards. If they know their route like the back of their hand, they'll have any easier time doing this.

*Are Conductors Threatened?*

Some say that the era of the conductor is coming to an end. Others claim that conductors will have a place on the American landscape for decades to come. There's a case to be made for either view point. It's important to consider the future of the profession and the economic climate surrounding railroads when deciding whether or not to go into this line of work.

We've already stated that conductors and engineers share some responsibilities. In fact, given that most engineers were once conductors, most engineers are trained in all of the tasks that conductors do. Some train companies have begun to wonder if this makes the conductor's role redundant. Train companies like Montana Rail Link have even gone so far as to eliminate conductors altogether. Their freight trains are run by an engineer with an assistant engineer.

Most train companies still employ conductors on their freight trains. In fact, many of them are contractually obligated to do so. Union contracts with these companies insist that every train be accompanies by at least one conductor. Said contracts would have to be renegotiated before conductors could be removed from freight train crews.

Such changes are opposed not only by the conductors' union, but also by the Brotherhood of Locomotive Engineers. Although engineers are technically capable of doing all that conductors do, they also have very important jobs of their own. The engineers' union has expressed concern over having engineers operate alone and have to split their attention. They fear it would be unsafe for engineers to concern themselves with anything else when involved in safely operating a train. Of course, this attitude doesn't apply to passenger train conductors, whose responsibilities require them to move about train cabins, something the engineer would be incapable of doing. However, their role may be threatened by fluctuating passenger volume. While passenger train conductors are certainly not going anywhere, they may find themselves few and far between in times of low demand.

*Future of Rail Jobs*

Conductors aren't the only ones who worry about finding themselves on the chopping block. The popularity and success of the railroad industry has waxed and waned over the decades, sometimes enjoying a heyday while other times falling fast toward obsoleteness. The question is where is it now and where is it headed. The answer varies dramatically depending on who you ask.

Some people insist that rail jobs are threatened because there's less demand for train transport (freight or otherwise). They say that train companies have been scaling back crews and even laying off good people. Those that are left urge others to avoid this industry altogether, warning that there's no future in it. But is this true? Maybe five years ago, or ten. At least that's what railroad supporters might say. They claim that these days freight travel is a faster and cheaper option for transporting goods across the country than trucking. Freight trains don't have to employ as many people per ton of goods than trucking companies do, making it easier for them to offer better prices. What's more, freight is a safer option than trucks. A moment's distraction on an eight-hour train ride is not nearly as likely to result in the disaster that a moment's distraction on the highway might. As more and more companies realize this and move to freight travel, more and more freight trains are being put into motion.

Freight trains are also more popular for bulk cargo like steel, iron, and coal. These tend to be exceedingly heavy, making it a challenge to transport them via truck. On the other side of that coin, however, is the fact that some bulk commodities (like coal) are no longer as strong as they used to be. As these industries flag, so too does the freight industry.

Finally, advances in technology have helped the passenger railroad industry as well. As trains get faster, they become more competitive with other modes of transportation. Suddenly Americans might prefer to take an Amtrak train from Boston to New York or from Denver to San Francisco over flying. What's more, as trains get stronger, they are able to hold greater numbers of people, making train travel more affordable than flying or, in some cases, even paying for gas.

In 2014, Forbes published an article claiming that the railroad industry was experiencing a new heyday for many of the aforementioned reasons. They stated that industrywide revenue, that was 67.7 billion dollars in 2009, had increased nineteen percent to 80.6 billion dollars by 2014. According to Forbes, and many others, this boom is a direct result of technological advances that have helped make the railroad relevant again. This is good news for prospective conductors, as the increased revenue also lead to 10,000 new jobs.

However increased demand is not the only boon working for the future of rail jobs. Job Monkey and similar sites that compile data on various industries claim that many railroad workers are due to retire in the coming years, opening up new opportunities for fledging conductors, brakemen, engineers, and more. This is also good news for those currently working their way up the ranks who may be frustrated by the heavy importance placed on seniority.

*What Does This Mean for Me?*

It means that, whether you are looking to stay a conductor for years to come or transition quickly into an engineering role, you've got a place at the railroad. Between healthy unions and a steadily increasing demand, conductors should be around for quite some time.

# A Case Study: The Long Island Rail Road

Since my son Bryan is approximately four years into his conductor career with the Long Island Railroad, I reasoned that the LIRR would be an appropriate case study regarding the conductor hiring process.

The Long Island Railroad is a commuter rail system in southeastern New York, stretching from Manhattan to the eastern tip of Suffolk County on Long Island. With an average weekday ridership of 337,800 passengers in 2014, it is the busiest commuter railroad in North America. It is also one of the world's few commuter systems that runs 24 hours a day, 7 days a week, year-round. It is publicly owned by the Metropolitan Transportation Authority, who refer to it as MTA Long Island Rail Road. There are 124 stations, and more than 700 miles of track, on its two lines to the two forks of the island and eight major branches, with the passenger railroad system totaling 319 miles route.

*Hiring Practices*

The hiring of LIRR assistant conductors and conductors is carried out through an extensive assistant conductor pre-hire qualification process and training program. The conductor position is a promotional opportunity from the LIRR assistant conductor position. Applicants from outside the agency must apply for and serve as LIRR assistant conductors prior to promotion to conductor.

Individuals, who have conductor experience from other railroads, are hired as assistant conductors and are required to go through the assistant conductor training program. This is due to the information that is unique to the LIRR equipment, revenue, operating rules and physical characteristics. There are also union requirements regarding conductor seniority and roster position.

The minimum qualifications for assistant conductors and conductors are the same, the most definitive being a High School Diploma, with cash handling experience, customer service experience, and some college preferred. Additionally, conductors must have a minimum of two years of LIRR assistant conductor experience prior to appointment. Based upon need, the LIRR management requires assistant conductors to qualify as conductors. Conductors are required to pass additional written and practical tests. The LIRR Human Resources and Transportation Departments generally hold two or three open house meetings for assistant conductor applicants every year. Applicants are recruited through notices in local metropolitan area newspapers, the MTA and LIRR websites and other LIRR employees. Approximately 40 percent of those eventually hired are LIRR employee referrals.

The three hour open house sessions provide an overview of the responsibilities and training required of an assistant conductor. A video, "Miracle on 34th Street", is shown to applicants to introduce the totality of railroad operations. LIRR staff conducts fifteen minute exploratory interviews with applicants to go over individual resumes and applications as well as experience with customer service and cash handling. The staff also administers math and vocabulary tests. Of the 400 invitations sent to attend the open house, approximately 260 people attend. Of those attendees, roughly two hundred pass the pre-qualification requirements.

The second step of the application process is a general background check to verify application information, determine employment reliability and credit worthiness. About 100 of the applicants usually make it through this step.

For the third step, applicants are invited to a formal interview with a panel of two LIRR managers, one from Human Resources and one from the Transportation Department. Real life scenarios are presented to applicants to see how they would handle certain situations. Applicants are also asked to write an essay about how they would relate to customers. Roughly 60 out of the 100 applicants are invited to proceed.

The fourth step is a one day course, held on a Saturday, which includes signal and railroad rules. Applicants are asked to return a month later to take a test on the material.

The next step in the process was recently changed. Previously, the fifth step began the first phase of the formal assistant conductor training program, administered jointly by the Human Resources and Transportation Departments. Human Resources Department staff presented the material and Transportation Department staff administered tests and regularly evaluated the trainees. Applicants were not paid to take the course, which was held two weekday evenings and a full day on Saturdays over 15 or 16 weeks. The course covered ticket selling, the book of rules, the air brake and information required under federal and state government regulations, such as CPR training and emergency preparation. The importance of customer service was stressed throughout the course. Trainees who completed the course successfully were hired by LIRR and continued with phase two of the training (four weeks). Presently, all training is conducted after hiring. Trainees are required to pass a physical exam prior to hiring, before commencing an eleven week full time training course as salaried employees. As illustrated in the prior hiring steps, a pool of approximately 400 candidates ends up being whittled down to the normal sized training class of 25 trainees.

*Supervisory Structure and Job Responsibilities*

As of January, 2017 there are approximately 300 assistant conductors and 1050 conductors working at the LIRR. Assistant conductors and conductors are directly supervised by 28 transportation managers.

On-train personnel, exclusive of the engineer, always include a qualified conductor and an assistant conductor as a minimum. The assistant conductor on-train position may be filled by an assistant conductor or qualified conductor. Assistant conductors and qualified conductors may also be assigned to work as ticket collectors on specific trains or specific trip segments, as needed. In this capacity, assistant conductors or conductors collect passenger tickets and do not function as conductors. All three train service personnel positions are directly supervised by the 28 transportation managers. The transportation managers, a first level non-union management position, are responsible for the day to day transportation operations. In that capacity, the transportation managers supervise a range of personnel who are involved in these operations including: on-train personnel, assistant conductors, conductors, ticket collectors and engineers, in addition to assistant station masters, ushers and yardmasters. Each manager is typically responsible for 38 employees.

Transportation managers must have successfully completed a one year Transportation Department Management Development - Transportation Supervisor Training Program. This program is open to most Transportation Department operating personnel, but spaces in the program are available only as the need to replace transportation managers arises. This mentor type program prepares trainees for management and includes a rotation through different terminals and departments.

The Transportation Department managers report to nine lead field, functional or terminal managers, who manage a combination of field activities, key terminal locations and functional areas, such as the following of railroad rules. The nine lead transportation managers report to ten superintendents.

*Training Practices*

The training for assistant conductors covers ticket selling, the book of rules, the air brake and information required under federal and state government regulations, such as CPR training and emergency preparation. Additionally, several weeks are devoted to the train equipment. The course also includes one week of on-the-job mentored field experience supervised by the Transportation Department.

The assistant conductor training program is taught by former LIRR conductors, who are certified technical trainers. The trainers are required to have been qualified LIRR conductors and have earned a technical training certificate.

While the eleven week training segment is mostly lecture, it is broken up with field visits, practical exercise and role playing techniques. The first two days of the training consist of the general orientation overview of LIRR given to all new Rail Road employees including diversity issues, equal employment opportunities and the federally mandated "Right to Know" safety training. The remaining weeks include a visit to the Ticket Receiver Department (which functions as the liaison between the Finance and Transportation Departments), familiarization with Penn Station and the West Side Yard, and discussions and lectures about fare collection. Also included is a focus on tunnel evacuation and meeting with the unions, and also a day is spent coupling and uncoupling the cars and engines and working the switches. Additional days are spent on revenue service trains for on-the-job training with the instructors.

Assistant conductor trainees spend the next week one-on-one with a mentor. Mentors are qualified conductors who have been chosen by the Transportation Department because of their consistent outstanding performance and absence of customer complaints. Many of the mentors who participate have benefited from the mentor program themselves. The mentors are sent a letter outlining the expectations for the day and a check list of items to be covered per given day. The mentor training focuses on making announcements on board trains, learning how to provide quality customer service and practicing operating procedures. Mentors evaluate trainee performance according to a standard evaluation form. After the week of mentoring, the trainees have four additional days of training, which cover crime intervention, familiarization with all the terminals and meeting the supervision staff, selecting job assignments and graduation. Final exams are administered to the trainees by the rules examiners for each portion of the training: the rules, air brake, ticket and passenger train emergency preparedness (PTEP). Those trainees who fail any of the required tests are immediately terminated. Bryan's class lost three trainees throughout the training program, including a young lady who failed the test on the final day of class, which also happened to be her birthday.

The customer service portion of the training, integrated throughout the program, makes use of videos and role playing techniques to focus on topics, such as making announcements and helping customers to board and alight from the train. The final day of training concludes with a graduation ceremony and a motivational lecture about the importance of safety, communication and customer service.

All assistant conductors are required to qualify as conductors, between two and five years after initial assistant conductor qualification, as is stipulated in their labor agreement with LIRR.

Failure to qualify as a conductor will result in the termination of employment.

The conductor training program consists of 21 days of training including topics, such as rules, physical characteristics, air brake and operating procedures in New York's Penn Station. An additional five days are allocated to conductors for on-the-job mentor training to allow candidates to overcome their fear of being in charge and become comfortable with their new responsibilities. The mentor uses a standard form to evaluate the trainee's performance.

Tests are given in each one of the specific topic areas. The assistant conductor trainee must pass all of the tests prior to promotion to conductor. I can tell you first hand from my experience coaching Bryan through his qualification, these tests are brutal. The amount of material that has to be memorized is staggering, and there is much more material included than in the eleven week initial training class. For example, the qualification includes a test on the physical characteristics of the railroad in which the assistant conductor has to draw the entire railroad. That's right – every track, station, switch, and siding over the entire 700 miles of track. At least during the qualification testing it is not one and done. Assistant conductors can take a failed test six times before facing termination. However, upon the third failure the assistant conductor is taken out of service and takes the exam once each month until he passes or is terminated.

# Conclusion: A Noble History

At the beginning of this book, we asked you to picture the conductor and we all imagined Ringo Starr at Shining Time Station. Let's picture the conductor again. Only this time, let's imagine the friendly late-night conductor on the Polar Express. It seems fitting that a story which seeks to combine comfort, coziness, and adventure should feature such a figure. In fact, it seems doubly fitting that railroad conductors are featured so prominently in children's stories overall. We have also seen, however, that the conductor is so much more than a child's fairytale character. The conductor ensures safe passage for millions of riders each day and is the first line of response to emergency situations. We must ask ourselves what it is about the railroad that speaks to the sense of adventure, wonderment, and security in all of us. If you are drawn to a career as a railroad conductor, then perhaps you already know. The life of the railroad conductor is a dichotomous one, balancing freedom and responsibility. It is a job for people who know how to wait and how to command respect. In short, it is both a symbol of a functioning society and a stark contrast to the regular 9-5 jobs that we, as members of such a society, are expected to get.

If train conducting is not your path, you should still be able to respect the role and what it has come to symbolize for children, America, and anyone who dreams of something different. Personally, I have developed the utmost respect for the job that these dedicated men and women perform, and even more respect for the efforts and studying they put forth to obtain their qualifications. Finally, I am indebted to the conductor profession and the railroad because it has provided me with a new interest and hobby. There are limits, however, to my newfound interest. You won't find me standing on a station platform taking videos of the various locomotives. Not yet, at least.

# Glossary of Terms

The following terms and companies are specific to the railroad industry and appear in the text. These do not include any of the jargon discussed in Chapter Nine.

Amtrak – A North American passenger train company that is also known as the National Railroad Passenger Corporation.

Assistant conductor – Answers to the conductor and helps perform various duties on passenger trains.

Automatic blocking signal (ABS) – An automated system for communicating on railroads that divides a rail line into blocks and controls rail movement to prevent collisions.

Ballast – Large rocky material that makes up the bedding around tracks.

Bidding – The process for appealing for a new position within a rail company. Conductors might bid on spaces in an engineering training program.

BNSF Railway – Second largest freight railroad company in North America

Brakeman – Men who would apply train brakes in the early days of the railroad. Now the position is a catch all position with various duties.

Brotherhood of Locomotive Engineers – The first train union. It was founded in 1863. These days it is sometimes known as the Brotherhood of Locomotive Engineers and Trainmen.

Brotherhood of Locomotive Firemen and Enginemen – A union founded in 1906. It has since merged with other rail unions and its constituents represented by SMART.

Brotherhood of Railway Brakemen – A union founded in 1883. It has since merged with other rail unions and its constituents represented by SMART.

Brotherhood of Railway Carmen – A union founded in 1890. It now exists as a division of the Transportation Communication International Union (TCU)

Bulk cargo – Unpacked cargo. Typically, it is a resource like coal or steel.

Bulletins – Brief statements or announcements that may contain information pertinent to a trip.

Caboose – The last car of a freight train where the conductor would stay. These are no longer in use except in rare circumstances.

Cargo – The contents of a train (that which is being shipped).

Carmen – Essentially the maintenance men of the train. They are responsible for caring for the train cars.

Central Pacific Route – One of the two transcontinental railroad routes. This one stretched from

Sacramento, CA to Promontory Summit, Utah.

Common carrier – A rail line that can be rented out to private or public entities as opposed to being controlled and used by a single entity.

Conditional employment – A promise of employment contingent on the fulfillment of certain requirements.

Conductor – The "captain" of a train who is responsible for the schedule and the safety of all passengers, crew, and cargo.

Consignment note – A contact that details the specific conditions and shipping information regarding specific cargo.

Coupling – Attaching rail cars.

CSX – Third largest freight railroad company in North America.

Deadhead – A train employee who is hitching a ride on a train to get to a certain destination (usually home) without working on the train.

Derail – When a train detaches from the tracks.

Dispatcher – Facilitates train movement by communicating with conductors and, sometimes, engineers from a remote location.

Employment Fair – A gathering where prospective railroad employees are assessed and interviewed before being offered positions with the rail company.

Engineer – Also called engine drivers and train operators. They drive the train.

Fireman – Manage the energy needs of diesel trains and the few steam engines still left in existence. Historically, they shoveled coal to keep steam engines running.

Freight train – Dedicated to the transport of non-human cargo from one point to another.

Guards – Name for conductors in India.

Hang test – Requires conductor trainees to hang onto the side of a rail car while signaling or performing some other action with their free hand.

International Association of Sheet, Metal, Air, Rail and Transportation Workers (SMART) – The modern iteration of the conductor union. It serves other railroad jobs as well.

Locomotive – The front car of a train that contains the engine and powers the rest of the vehicle.

On-call – A work situation where employees must await phone calls to tell them when and where to show up for shifts.

On-the-job training – Training, typically paid, that involves working in the desired position as you learn more about it.

Order of Railway Conductors – The original conductor's union. It was established in 1868 and has since merged with other rail unions and its constituents represented by SMART.

Pacific Railroad Act – An 1862 act that offered funding to build a railroad clear across the country to the Pacific Ocean. This resulted in the Transcontinental Railroad.

Passenger train – Dedicated to the transport of people from one location to another.

Pre-employment exams – Various tests required by railroad companies before they will consider applicants for employment.

Rail line – A specific track connecting two destinations.

Railroad – A set of tracks along which trains run. This is the American version of railway.

Railway – A set of tracks along which trains run. This term is the international version of railroad.

Reciprocating engine – Also called a piston engine. This type of engine uses a repetitive motion to create pressure which is then translated into movement.

Route – The schedule of locations that a conductor must direct a train through.

Shift work – Work that doesn't adhere to traditional daytime schedules and often results in inconsistent and inconvenient hours.

Shinkansen – Japanese high speed rail lines.

Shisa kanko – A communication system used by train crew in Japan.

Shunting – Shifting and switching of train cars. In America this is called switching.

Signals – Nonverbal forms of communication utilized by train crew (primarily conductors and engineers).

Station – Ports that trains drive in and out of to deposit or pick up goods or passengers.

Steam engine – An engine that runs off of steam, the invention of which help give birth to locomotives.

Subway – Trains that operate within urban centers, typically underground.

Switches – Located where tracks split, these devices switch which track the main one is leading to, sending trains either left or right.

Switching – Shifting and switching of train cars. In Europe this is called shunting.

Switch list – Keeps track of changing cars on a train. It is a small pocket document that is used in lieu of bringing a stack of waybills out of the train.

Timetables – A schedule of departures and arrivals for the train.

Tour of duty – More than just a shift, this term encompasses a shift, an overnight stay in a motel, and the trip home.

Track obstruction – Anything located on the tracks that might threaten to damage or derail the train.

Tracks – Steel rails upon which trains run.

Train – A large collection of cars containing goods or passengers with a locomotive or train engine at the front and, historically, a caboose at the back.

Transcontinental – Traversing an entire continent. The transcontinental railroad went from the west coast nearly to the east coast.

Trolley – A small car that runs on tracks and is connected to an overhead wire. These are typically found in cities.

Uncoupling – Detaching rail cars.

Union Pacific Route - One of the two transcontinental railroad routes. This one stretched from Bluffs, IA to Promontory Summit, Utah.

Union Pacific Railroad – Largest freight railroad company in North America.

Unions – an organized collection of workers within a specific industry that works to protect their rights, safety, and other interests.

United Transportation Union – A large union that represents railroad, bus, airline, and other workers.

Wagonways – Predecessors to the modern railroad, these grooves in stone paths helped ancient Greeks move carts.

Waybill – Paperwork that deals with the contents of various train cars. Each car will have its own waybill.

Wayside signal – A signal (like a traffic light) along the train tracks that helps direct train movement.

Work orders – Provides overall information about the current tour of duty.

# Bibliography

"Amtrak - Benefits." *Amtrak - Benefits*. N.p., n.d. Web. 11 Nov.

    2016.

"Automatic Block Signaling." *Wikipedia*. Wikimedia Foundation,

    n.d. Web. 04 Oct. 2016.

"Best Foot Forward, Training Frontline Personnel to Provide Quality

    Customer Sevice." *Permanent Citizens Advisory Committee*

    *to the MTA,* Nov. 2003.

"China: How to Travel by Train in China." *China: How to Travel by*

    *Train in China - TripAdvisor*. N.p., n.d. Web. 05 Nov. 2016.

"Conductor (rail)." *Wikipedia*. Wikimedia Foundation, n.d. Web. 30

    Sept. 2016.

"Crude Oil Transportation: A Timeline of Failure." *Riverkeeper*.

    N.p., n.d. Web. 03 Jan. 2017.

"CSX Careers." *CSX*. N.p., n.d. Web. 10 Nov. 2016.

"Early American Railroads." *Ushistory.org*. Independence Hall

    Association, n.d. Web. 04 Oct. 2016.

"The First Railroad Accident." *History.com*. A&E Television

    Networks, n.d. Web. 03 Jan. 2017.

"First Transcontinental Railroad." *Wikipedia*. Wikimedia

Foundation, n.d. Web. 04 Oct. 2016.

"The Five Deadliest Train Derailments in US History." *The Daily

Beast*. The Daily Beast Company, n.d. Web. 02 Jan. 2017.

"Freight Conductor." *Union Pacific: Building America*. N.p., n.d.

Web. 10 Nov. 2016.

"The Freight Train Conductor." *American-Rails.com*. N.p., n.d. Web.

28 Oct. 2016.

Geiling, Natasha. "'They Did Everything They Could Have Done':

The Tragedy Of The Oregon Oil Derailment."

*ThinkProgress*. ThinkProgress, 08 Aug. 2016. Web. 03 Jan.

2017.

"History of Rail Transport." *Wikipedia*. Wikimedia Foundation, n.d.

Web. 30 Sept. 2016.

"History of the U.S. Telegraph Industry." *EHnet*. N.p., n.d. Web. 04

Oct. 2016.

"International Brotherhood of Teamsters." *Wikipedia*. Wikimedia

Foundation, n.d. Web. 04 Jan. 2017.

Lam, Bourree. "Working on the Railroad." *The Atlantic*. Atlantic

Media Company, n.d. Web. 14 Dec. 2016.

"Lives on the Railroad: Railroad Conductor." *America on the Move.*

N.p., n.d. Web. 24 Oct. 2016.

Lynch, Dennis. "What Is The Difference Between A Train

Conductor And An Engineer?" *International Business Times.*

N.p., 13 May 2015. Web. 13 Dec. 2016.

"More Than Meets the Eye: A Conductor's Uniform."

*Blog.amtrak.com.* N.p., 22 Feb. 2013. Web. 10 Nov. 2016.

Northern Pacific Railway Company. *Rules and Specifications*

*Regarding Uniforms of Employees.* N.p.: n.p., 1913.

*Washington State Historical Society.* Washington State

Historical Society. Web. 10 Nov. 2016.

Orin, Andy. "Career Spotlight: What I Do as a Train Engineer."

*Lifehacker.* Lifehacker.com, 03 Feb. 2015. Web. 13 Dec.

2016.

"Pacific Railroad Acts." *Wikipedia.* Wikimedia Foundation, n.d.

Web. 06 Oct. 2016.

"The Pacific Railway - A Brief History of Building the

Transcontinental Railroad." *The Transcontinental Railroad.*

N.p., n.d. Web. 04 Oct. 2016.

"Rail Transport Trends | Future of Jobs with Railroads." *JobMonkey*. N.p., n.d. Web. 13 Dec. 2016.

"Railroad Conductor Salaries." *Glassdoor*. N.p., n.d. Web. 10 Nov. 2016.

"Railroad Fireman." *Inside Jobs*. N.p., n.d. Web. 11 Oct. 2016.

"Railroad Glossary And Terms." *American-Rails.com*. N.p., n.d. Web. 06 Jan. 2017.

"Railroad History, An Overview Of The Past." *American-Rails.com*. N.p., n.d. Web. 30 Sept. 2016.

"Railroad Language -- Lingo -- Dictionary." *Catskill Archive*. N.p., n.d. Web. 06 Jan. 2017.

"Railroad Slang." *Alaska Rails*. N.p., n.d. Web. 06 Jan. 2017.

"Railroad Uniforms." *Railroad Uniforms - Railroadiana Online*. N.p., n.d. Web. 10 Nov. 2016.

"Reciprocating Engine." *Wikipedia*. Wikimedia Foundation, n.d. Web. 30 Sept. 2016.

"RWU's History." *Railroad Workers United*. N.p., n.d. Web. 04 Jan. 2017.

Schubert, Holly. "What Does a Railroad Conductor Do?" *The Balance*. N.p., n.d. Web. 24 Oct. 2016.

Skye, Stephen. "The Life of a Brakeman." *The Neversink Valley Museum of History Innovation.* N.p., n.d. Web. 11 Oct. 2016.

Thale, Christopher. "Railroad Workers." *Encyclopedia of Chicago.* N.p., n.d. Web. 24 Oct. 2016.

"Timeline of United States Railway History." *Wikipedia.* Wikimedia Foundation, n.d. Web. 30 Sept. 2016.

"Train Conductor: Train Guard, Train Crew." *National Careers Service.* N.p., n.d. Web. 10 Nov. 2016.

"Transcontinental Railroad." *History.com.* A&E Television Networks, 2010. Web. 07 Oct. 2016.

Vartabedian, Ralph. "Why Are so Many Oil Trains Crashing? Track Problems May Be to Blame." *Los Angeles Times.* Los Angeles Times, n.d. Web. 30 Dec. 2016.

"Who Are The Teamsters?" *Teamsters.* N.p., n.d. Web. 04 Jan. 2017.

Made in the USA
Columbia, SC
17 May 2020